Springer Series in Materials Science

Volume 277

The Springer Series in Materials Science covers the complete spectrum of materials physics, including fundamental principles, physical properties, materials theory and design. Recognizing the increasing importance of materials science in future device technologies, the book titles in this series reflect the state-of-the-art in understanding and controlling the structure and properties of all important classes of materials.

More information about this series at http://www.springer.com/series/856

Suprakas Sinha Ray
Editor

Processing of Polymer-based Nanocomposites

Introduction

Springer

Editor
Suprakas Sinha Ray
DST-CSIR National Centre for
Nanostructured Materials
Council for Scientific and Industrial
Research
Pretoria, South Africa

and

Department of Applied Chemistry
University of Johannesburg
Johannesburg, South Africa

ISSN 0933-033X ISSN 2196-2812 (electronic)
Springer Series in Materials Science
ISBN 978-3-319-97778-2 ISBN 978-3-319-97779-9 (eBook)
https://doi.org/10.1007/978-3-319-97779-9

Library of Congress Control Number: 2018950820

This Springer imprint is published by the registered company Springer Nature Switzerland AG
The registered company address is: Gewerbestrasse 11, 6330 Cham, Switzerland

Preface

Over the last few years, "nanomaterial" and "nanotechnology" have become well-known terms, not only among scientists, engineers, fashion designers, and architects, but also the general public. Owing to their extraordinary and unexpected behavior, nanomaterials have gained tremendous attention in fields such as automotive, electronics, aerospace, health care, and biomedical and have significant potential for many modern advanced technological applications. In this direction, a great deal of research and development effort has emerged around the hybrid organic–inorganic systems, and, in particular, attention has been given to those in which nanofillers (or nanoparticles, NPs) are dispersed in a polymer matrix. This class of materials is called polymer nanocomposites (PNCs) and shows unique value-added properties that are completely absent in neat matrices and conventional composites.

NPs can be made from a wide range of materials, the most common being layered silicates, carbon nanotubes, graphene or graphite oxide, and metal oxides. Over the last two decades, various types of NPs have been used for the preparation of PNCs with almost all types of polymers and polymer blends. However, layered silicates, carbon nanotubes, and graphene-containing PNCs have attracted in current advanced materials research, because these NPs can remarkably enhance the inherent properties of neat polymers after PNCs formation.

Control of the dispersion of NPs in polymer matrices and the interactions between NPs and macromolecular chains are necessary to explore PNCs full potential for a wide range of applications. This can be achieved by proper surface functionalization and modification of the NPs, which enhance their interaction with matrix. Functionalization of NPs consequently influences the colloidal stability, dispersion, and controlled assembly of NPs.

The performance of a PNC is dictated by three main factors: (i) the inherent properties of the components; (ii) interfacial interactions; and (iii) structure of the PNCs. The structure of a PNC depends on the dispersion and distribution of the NPs in the polymer matrix. However, improving the dispersion by mechanical means or via chemical bonding can influence the properties of the obtained PNCs.

Therefore, elucidating the dispersion and distribution characteristics and the associated mechanisms is important and can allow prediction of the final properties.

Over the past few years, great deals of advancements have been made in PNCs research and development, and the dispersion of the NPs in the polymer matrix remains a key challenge to their widespread application. Undoubtedly, to maximize the interfacial area in the PNCs, the dispersion should be on the scale of the individual NPs, otherwise aggregation results in a lower specific surface area, and microcomposites are ultimately formed. A large portion of this book is dedicated to reviewing methods for manipulating the interface, as well as the kinetic aspects of the dispersion of nanoparticles in polymer matrices, for example, controlling extrusion parameters during melt-processing.

Another pressing matter is the control of the structural morphology of the PNCs to achieve time-independent equilibrium. Often, further processing, for example, extrusion and then injection molding, results in a different dynamic equilibrium, which makes it difficult to tune the emergent properties. A similar complex challenge involves scaling-up preparation of the PNCs to industrially viable quantities, especially for solvent-based systems. To secure economic and other societal impacts, sufficient volumes of NPs and PNCs need to be manufactured in order to be transformed into market-ready products. Most research and development laboratories around the world focus on processes with tightly controlled laboratory environments which are not necessarily appropriate for safe, reliable, effective, and affordable large-scale production. Funding to address these challenges is still required. In addition, as with other new technologies, societal aspects need to be considered in parallel with the development of the technology. Consumer perceptions regarding the risks of certain nanoparticles, whether proven or not, need to be addressed to ensure a smoother uptake of the technology.

In summary, the field of PNCs has shown disparate results, successes as well as challenges. Much work remains to be done in fundamental research to achieve better control of the desired properties, for example, processing PNCs to achieve the desired level of dispersion of the NPs. This book focusses the effect of process variables on the structural evolution of polymer composites and the production of either micro- or nanocomposites, as well as the effect of the dispersion state on the final properties at a fundamental level.

Processing these PNCs usually requires special attention as the resultant structures on the micro- and nanolevel are directly influenced by the polymer/NP chemistry and processing strategy, among others. The structure then affects the properties of the resultant composite materials. This book is structured into two volumes. The first volume introduces readers to nanomaterials and PNCs processing. After defining NPs and PNCs and discussing environmental aspects, Chap. 2 focuses on the synthesis and functionalization of nanomaterials with applications in PNC technology. A brief overview on nanoclay and nanoclay-containing PNC formation is provided in Chap. 3. Chapter 4 provides an overview of the PNCs structural elucidation techniques, such as X-ray diffraction and scattering, microscopy and spectroscopy, nuclear magnetic resonance, Fourier transform infrared spectroscopy and microscopy, and rheology. The last chapter provides an overview on how melt-processing strategy impacts structure and mechanical properties of polymer nanocomposites by taking layered silicate-containing polypropylene nanocomposite as a model system.

The second volume focuses heavily on the processing technologies and strategies and extensively addresses the processing–structure–property–performance relationships in a wide range of polymer nanocomposites, such as commodity polymers (Chap. 1), engineering polymers (Chap. 2), elastomers (Chap. 3), thermosets (Chap. 4), biopolymers (Chap. 5), polymer blends (Chap. 6), and electrospun polymer (Chap. 7). The important role played by NPs in polymer blends structures in particular is illustrated.

This two-volume book is useful to undergraduate and postgraduate students (polymer engineering, materials science and engineering, chemical and process engineering), as well as research and development personnel, engineers, and material scientists.

Finally, I express my sincerest appreciation to all authors for their valuable contributions as well as reviewers for their critical evaluation of the proposal and manuscripts. I also thank all authors and publishers for their permission to reproduce their published works. My special thanks go to Kirsten Theunissen, Aldo Rampioni, Editor, and Production Manager at Springer Nature for their suggestions, cooperation, and advice during the various stages of manuscripts preparation, organization, and production of this book. The financial support from the Council for Scientific and Industrial Research, the Department of Science and Technology, and the University of Johannesburg is highly appreciated. Last but not least, I would like to thank my wife Prof. Jayita Bandyopadhyay and my son Master Shariqsrijon Sinha Ray, for their tireless support and encouragement.

Pretoria/Johannesburg, South Africa Suprakas Sinha Ray

Contents

Contributors

Jayita Bandyopadhyay DST-CSIR National Centre for Nanostructured Materials, Council for Scientific and Industrial Research, Pretoria, South Africa

Neeraj Kumar DST-CSIR National Centre for Nanostructured Materials, Council for Scientific and Industrial Research, Pretoria, South Africa

Dimakatso Morajane DST-CSIR National Centre for Nanostructured Materials, Council for Scientific and Industrial Research, Pretoria, South Africa; Department of Applied Chemistry, University of Johannesburg, Doornfontein, Johannesburg, South Africa

Vincent Ojijo DST-CSIR National Centre for Nanostructured Materials, Council for Scientific and Industrial Research, Pretoria, South Africa

Suprakas Sinha Ray DST-CSIR National Centre for Nanostructured Materials, Council for Scientific and Industrial Research, Pretoria, South Africa; Department of Applied Chemistry, University of Johannesburg, Doornfontein, Johannesburg, South Africa

Chapter 1
Introduction to Nanomaterials and Polymer Nanocomposite Processing

Vincent Ojijo and Suprakas Sinha Ray

Abstract Polymer nanocomposites are a relatively new class of material containing components with at least one dimension less than 100 nm. This chapter provides an introduction to these polymer nanocomposites and their relationship to microcomposites. The overall "nano-effect" in these advanced composites is briefly discussed. Research and development trends in the field of nanotechnology (including nanocomposites) are also presented. Finally, the chapter discusses current views on pertinent issues related to nanotechnology, including the environment, safety, health, and ethics.

1.1 Nanomaterials and Their Significance

It is generally accepted that nanomaterials are structures with at least one of their dimensions measuring below 100 nm. The European Union adopted an expanded definition that accounts for environmental, and health and safety issues, vide, *recommendation on definition of a nanomaterial* (*2011*/696/EU) [1]:

A nanomaterials is "*a natural, incidental or manufactured material containing particles, in an unbound state or as an aggregate or as an agglomerate and where, for 50% or more of the particles in the number size distribution, one or more external dimensions is in the size range 1 nm to 100 nm.*

V. Ojijo (✉) · S. Sinha Ray
DST-CSIR National Centre for Nanostructured Materials,
Council for Scientific and Industrial Research, Pretoria 0001, South Africa
e-mail: vojijo@csir.co.za; rsuprakas@csir.co.za

S. Sinha Ray
Department of Applied Chemistry, University of Johannesburg,
Doornfontein 2028, Johannesburg, South Africa
e-mail: ssinharay@uj.ac.za

S. Sinha Ray (ed.), *Processing of Polymer-based Nanocomposites*,
Springer Series in Materials Science 277,
https://doi.org/10.1007/978-3-319-97779-9_1

In specific cases and where warranted by concerns for the environment, health, safety or competitiveness the number size distribution threshold of 50% may be replaced by a threshold between 1 and 50%".

The National Nanotechnology Institute (NNI) definition of a nanometer includes a dimensional perspective:

"***A nanometer*** *is one-billionth of a meter. To illustrate, a sheet of paper is about 100,000 nm thick. Unusual physical, chemical, and biological properties can emerge in materials at the nanoscale. These properties may differ in important ways from the properties of bulk materials and single atoms or molecules*" [2].

The different emergent properties of nanomaterials compared to their bulk counterparts are attributed to their extremely high specific surface area and quantum confinement effects. The latter results in dramatic changes to the electronic and optical properties; controlling the size of the nanomaterial allows these properties to be tuned for the requirements of specific applications.

1.2 Overview of Nanotechnology

1.2.1 Definitions

Although examples of nanostructured materials are centuries old, Professor Norio Taniguchi of Tokyo Science University first coined the term '*nanotechnology*' at the 1st International Conference on Precision Engineering (ICPE) in 1974 [3]. Since then, the concept has been widely applied and attracted a huge amount of interest, mostly in research and development (R&D), but also for commercial applications. Nanotechnology is widely considered a key enabling technology and has opened new possibilities in the fields of medicine, electronics, sensors, catalysis, polymer nanocomposites, and others.

The National Nanotechnology Initiative (USA) defines nanotechnology, and outlines its basic tenets, as follows [4]:

"**Nanotechnology** *is the understanding and control of matter at dimensions between approximately 1 and 100 nm, where unique phenomena enable novel applications. Encompassing nanoscale science, engineering, and technology involves imaging, measuring, modelling, and manipulating matter at this length scale*.

1.2.2 Nanotechnology Funding

In both major and emerging economies around the world, there has been a substantial increase in spending for nanotechnology-related endeavors over the last two decades. Such funding was primarily directed to fundamental research, research infrastructure and instrumentation, and nanotechnology-enabled applications, devices, and systems. However, lately, more emphasis has been placed on the effects of nanomaterials on the environment, and human health and safety. For example, in the USA, the President's 2017 budget estimates provided for over USD 1.4 billion for the National Nanotechnology Initiative (NNI); since its inception in 2001, a total of approximately USD 24 billion has been channeled to the NNI to enable its various stakeholders to address the full complement of nanotechnology: R&D; infrastructure; human capital development; health, safety, and environment; commercialization; and other societal aspects [2].

A study funded by the U.S. National Nanotechnology Co-ordination Office and the U.S. National Science Foundation carried out by Lux Research Inc., revealed that the USA government leads the world in nanotechnology funding [5]. However, it was difficult to estimate nanotechnology spending by other governments as many have stopped the unified approach of tracking nanotechnology initiatives within their countries. In 2012 for instance, governments, corporations, and venture capitalists spent around USD 18.5 billion funding nanotechnology, where $\sim$36% of this was provided by the USA [5].

There was an initial drive towards central coordination of nanotechnology activities in the major economies. In the USA, this was achieved through the NNI formed in 2001 [4]. The NNI has been at the forefront in advancing nanotechnology-related activities, resulting in the USA taking an early lead in terms of the funding, quality, and number of R&D publications, and subsequent development of nano-enabled products. Following the establishment of the NNI, over 60 countries established national nanotechnology R&D programs between 2001 and 2004 [6]. Second to the USA, China, which often competes and collaborates with the USA in nanotechnology R&D, has lately ramped up activities in the nanotechnology field [7, 8]. The National Natural Science Foundation of China (NSFC) supports research in all nano-related areas, while a number of other institutions, including the Chinese Academy of Science (CAS), also play major roles [9, 10]. In Germany, there have been three federally coordinated plans outlining the federal government's new high-tech strategy. The first was the Nano Initiative Action Plan 2010 [11], which aimed to develop a standardized and multi-ministerial action framework for the coordination of objectives and methods for advancing nanotechnology. This was followed by Action Plan Nanotechnology 2015 [12] and Action Plan Nanotechnology 2020 [13]. In the Russian Federation, the Rusnano framework implements state policy for the development of the nanotechnology industry in Russia, and actively co-invests in projects based on nanotechnology that show meaningful social and economic potential [14].

1.2.3 Nanotechnology Outputs and Outcomes

Consistent with the increasing amount of funding for nanotechnology-related activities, research output, e.g.; journal articles, books, and conference papers, has

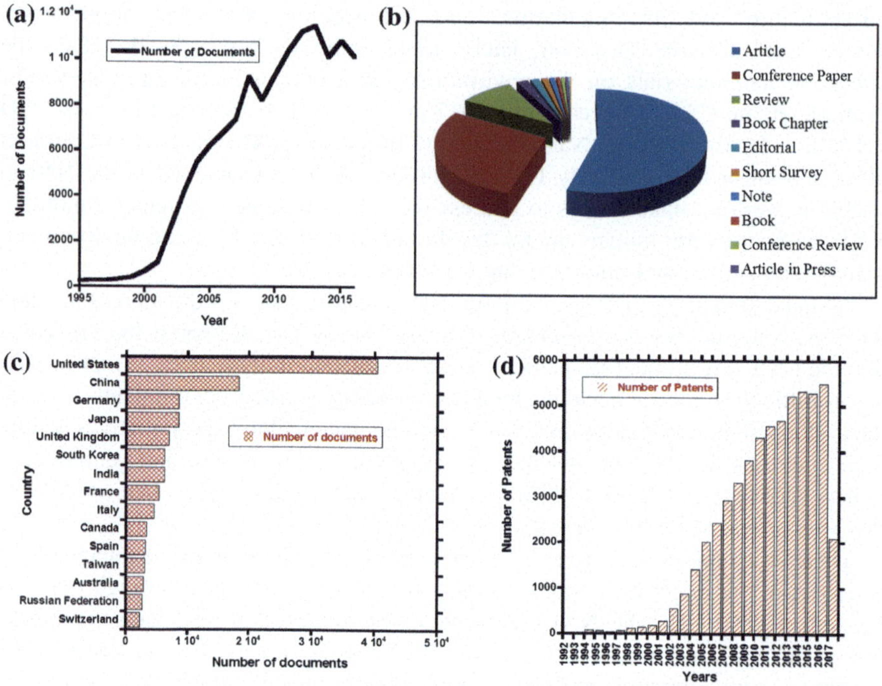

Fig. 1.1 **a** Number of research output documents per year; **b** document type; **c** the output of the top 15 countries; and **d** number of patents per year. *Source* Scopus, while the keyword for the search was; "Nanotechnology"

also been steadily increasing since 2000. This is depicted in Fig. 1.1a, while Fig. 1.1b shows the composition of these research outputs. With continued funding and interest from academia, along with increasing use of nanotechnology in products, increasing research output is expected to continue.

Figure 1.1c reveals that countries investing more in nanotechnology also generate the most research outputs, where the USA leads both. The number of patents has steadily increased since 2000 (see Fig. 1.1d), which is expected to be correlated with the number or value of nano-enabled products in the market. China is also playing an increasing role in nanotechnology R&D, which is expected to continue. The number of nanotechnology publications from China is second only to the USA; however, the citations per article are lower, indicating that the research quality may be lower than that of the USA and EU [7, 10]. Finally, revenue from nano-enabled products has increased and is forecast to continue to rise in the near future. In 2010, the revenue realized from nano-enabled products was in the range of USD 336 billion, and is forecast to reach over USD 3 trillion in 2018 [5].

1.3 Responsible Nanotechnology Development: Ethics, Society, Environment, and Health and Safety

Society is inherently a complex system. Within the society, competing values, interests, priorities and perceptions interfaces with science and technology. The sustainable adoption and exploitation of nanotechnology requires a balance between the use of the technology for social and financial gain and consideration of the ethics and effects on society [15]. Questions regarding how nanotechnology is introduced into society are expected to affect implementation of the technology; for example, whether stakeholders are involved in transparent decision making; whether legal, ethical, and social issues are addressed; and whether the technology addresses real and perceived needs of the stakeholders. Like any new technology, nanotechnology may offer both benefits and potential risks to humans and the environment during the development, use, and disposal of the technology.

As with many new technologies, the ethical considerations of nanotechnology have lagged behind the R&D [16]. Advocacy groups around the world have noticed this and are applying pressure for the safe application of nanotechnology. The USA government through the NNI recognized this matter and has adopted responsible development of nanotechnology as one of its four objectives. A component of this is consideration of the ethical, legal, and societal implications of nanotechnology, as well as its effects on the environment, and human health and safety [4, 17, 18]. Similar objectives have also been articulated by other countries. One of the strategic goals of the German federal government's Action Plan Nanotechnology 2020 [13] is: "*Ensuring responsible governance of nanotechnology as a contribution to sustainable development through accompanying risk research and communication*".

Bacchini [19] proposed that the ethical questions around nanotechnology are merely new instances of old ethical problems. However, other researchers take the opposite view, claiming that these ethical issues may not be very obvious, and could be emergent and variant, just as the scope of nanotechnology is broad [20]. We hold a similar opinion as Van de Poel [20], and believe that, due to the many different nanotechnologies, a single method of identifying and analyzing ethical issues for all nanotechnologies is not feasible; rather, each nanotechnology field should be considered on its own.

The responsible development and application of nanotechnology has gained currency and now efforts are being made to develop nanotechnology while concurrently considering its effect on the environmental, along with ethical, and health and safety implications. Various discussion forums regarding the ethics of the application of nanotechnology and nanomaterials have emerged, including the journals: "*Nanotoxicology*" and "*NanoEthics: Studies of New and Emerging Technologies*", founded in 2007. Comprehensive coverage of the ethical and social impacts of nanotechnology [21–27] and environmental and health and safety aspects [28–30] appears in other texts.

Responsible development of nanotechnology requires adherence to safe handling practices for nanoparticles (NPs) in order to protect the health of workers and users at all life cycle stages of the technology. Nanomaterials pose potential health and safety risks, primarily due to their small size, and could be more toxic than their bulk counterparts [28]. Schulte et al. [28] provided a good overview of the chronology of events in the main functional areas of toxicology, metrology, assessment of exposure, risk assessment and management, controls and personal protective devices, medical surveillance, and epidemiology, in order to identify potential health and safety risks for workers involved in nanotechnology value chain. For example, in the area of nanomaterial toxicology, emphasis has been placed on carbon nanotubes [31] as they are thought to promote interstitial fibrosis, while specific multi-walled carbon nanotubes have been shown to promote lung cancer [32]. In Fig. 1.2, Schulte et al. [28] illustrates the chronology of the toxicology of, mainly, carbon nanotubes.

The snapshot presented above reveals a growing interest in the nanotechnology field. There is increasing investments by both the government and private sectors in nanotechnology activities, while the returns in terms of revenue obtained from nano-enabled product are also growing. Responsible development of nanotechnology has also gained prominence and is now considered during the development, application, and disposal of nanoproducts. Among the nano-enabled products are

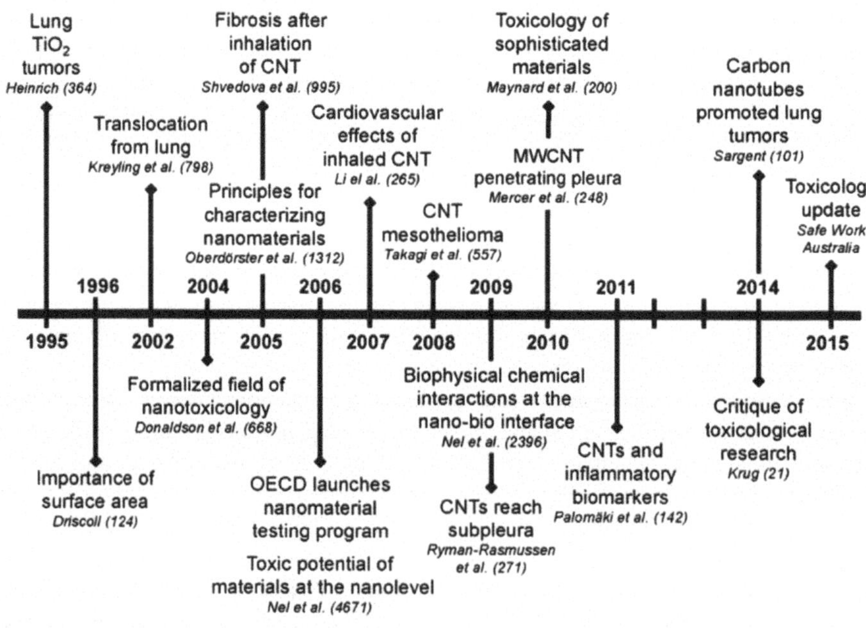

Fig. 1.2 Progress in nano-engineered materials—related toxicology. Reproduced with permission from [28]. Copyright 2016, Springer Verlag

those using polymer nanocomposites, which are the focus of this book. A brief introduction to polymer nanocomposites and the key differences between microcomposites and nanocomposites are presented in the next section.

1.4 Polymer Nanocomposites Versus Polymer Microcomposites

By definition, composite materials are bulk substances with two or more distinct constituents that are structurally complementary and have emergent properties not present in either of the components. In polymer nanocomposites, at least one of the components must have nanoscale dimensions, i.e., 1–100 nm. Traditional composites (microcomposites) contain components with a minimum dimension in the microscale. Although polymer nanocomposite research has been active for some time, it only gained prominence after ground-breaking work by the Toyota Central Research & Development Laboratory Inc. that showed that the addition of only 4.7 vol% of molecular (exfoliated) montmorillonite to nylon-6 enhanced the thermomechanical properties, such as the heat deflection temperature (increase of ~87 °C) [33]. Since then, many research outputs (considering publications) have been published, as shown in Fig. 1.3.

Likewise, the development of polymer nanocomposite products is on the rise, as shown in a report by BCC Research [34]. The global consumption of more than USD 1 billion in 2014 was forecast to increase to USD 4 billion in 2019, as shown in Fig. 1.4. Polymer nanocomposites are now used to make e.g., automobile parts, motorcycles, tennis rackets, baseball bats, and bicycles. The NNI gave the example of the use of carbon nanotubes, which combine the advantage of being lightweight with good conductivity (both electrical and thermal), to make next-generation motor cars, with the benefit of electromagnetic shielding and better thermal

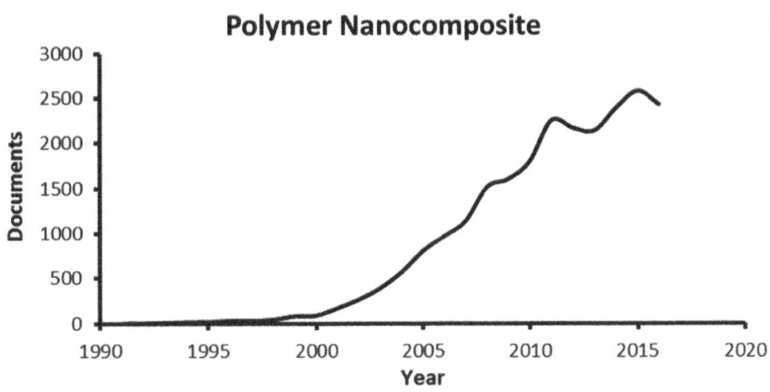

Fig. 1.3 Number of publications on nanocomposites *Source* Scopus, using the keywords 'Polymer Nanocomposite"

dissipation [4]. A well-established market is the tire sub-sector, where nanoparticles have been used to enhance certain properties, as highlighted by Kumar et al. [35] in a recent perspective. In the development of such products using polymer nanocomposites, various types of nanoparticles have been incorporated in various polymer matrices to harness emergent superior properties. These mainly include layered silicates [36–42], carbon nanotubes [43–45], and graphene [46–49]. When the nanoparticles are well dispersed, the resultant nanocomposites tend to display better emergent properties compared to either traditional microcomposites, or the inadvertent microcomposites resulting from poorly dispersed nanoparticles.

1.4.1 The "Nano-Effect"

The enhancement in the properties of polymer nanocomposites compared to microcomposites is thought to be due to different particle sizes of the dispersed particles. Polymer nanocomposites benefit from the synergy between such nanoparticles and polymer chains of the same size scale, as well as their high specific surface area (high interfacial area) compared to microcomposites [50]. Very recently, Cheng et al. [51] demonstrated that smaller nanoparticles (1.8 nm) affected the properties of polymers even more dramatically than larger nanoparticles (10–50 nm), in cases where the particle–polymer interactions are good. However, there is a critical particle size necessary to achieve property enhancements, where smaller particles can result in deterioration in e.g., mechanical properties. This could be due to the plasticization effect of extremely small particles, especially spherical ones.

To illustrate the enormous specific surface area of nanoparticles, consider a cube of 1 μm being split successively into parts until the dimensions are nanoscale (refer to Scheme 1.1). The surface area of the 1 μm cube is 6 μm^2, with a volume of 1 μm^3; hence, the specific surface area is 6 μm^{-1}. Cutting the cube into 8 smaller cubes of 0.5 μm gives a specific surface area of 12 μm^2, twice the original value. Further sub-division of the cubes into a billion smaller cubes, each with a length of

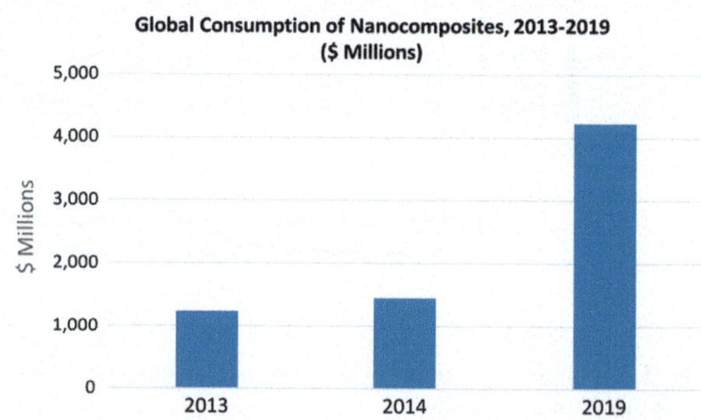

Fig. 1.4 Global consumption of nanocomposites, 2013–2019 (USD million). Reproduced with permission from [34]. Copyright 2014, BCC Research

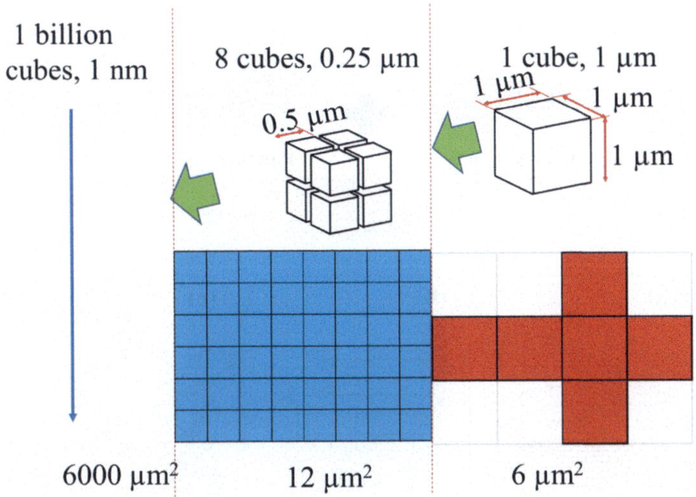

Scheme 1.1 Illustration of increased specific surface area upon change of dimensions from microscale to nanoscale

1 nm, gives a specific surface area of 6000 μm^2. This is 1000 times the specific surface area of the original microscale cube! This enormous specific surface area makes nanocomposites vastly different from their microcomposite counterparts. In particular, the volume fraction of the nanoparticles required to achieve modification of the polymer properties is therefore smaller than for microcomposites: ~1–5 vol % for layered silicate nanocomposites [39] and less than 1 vol% for carbon nanotube polymer nanocomposites [52]. When layered silicates are properly dispersed in a polymer matrix, they either form intercalated structures or are delaminated to give a nanocomposite, as shown in Fig. 1.5, where the thickness of the silicate is ~0.96 nm [36]. However, poor dispersion of the silicates in the polymer matrix results in a microcomposite, where the silicates remain as tactoids with a microscale thickness. Nanocomposites with delaminated or intercalated structures perform better than microcomposites [39].

Due to the important effects of the surface area of the nanoparticles in polymer matrices, the surface chemistry is an important consideration. For systems with good particle–polymer interactions, a boundary layer forms around the nanoparticles, and this dramatically alters the macroscopic properties of the resultant composite [53]. This boundary layer is depicted in Fig. 1.6, reproduced from Jouault et al. [53]. Different groups have reported different thicknesses for this immobilized layer. Thermogravimetric methods have shown a boundary layer of 1–5 nm thick [53], while others have reported larger values, in the range of the radius of gyration R_g of the chains [54].

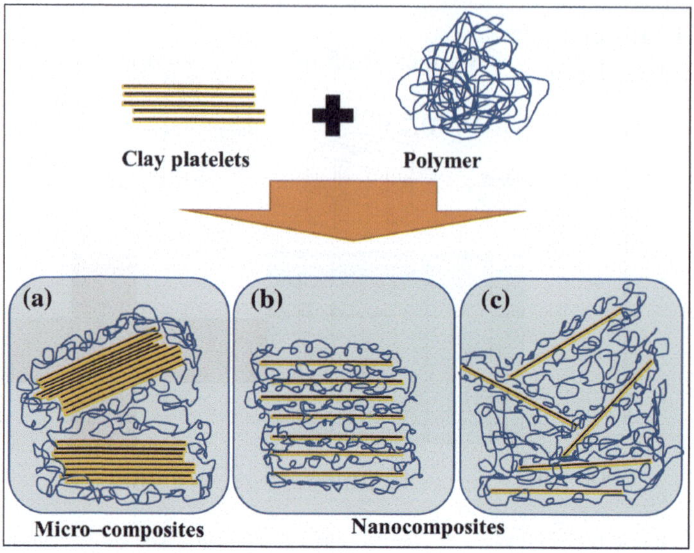

Fig. 1.5 Structures of layered silicate polymer composites: **a** micro-composites **b** intercalated nanocomposite; and **c** delaminated nanocomposites. Reproduced with permission from [36]. Copyright 1996, Wiley-VCH

Fig. 1.6 Depiction of the bound layer of polymer on silica particle. Reproduced with permission Adopted from [53]. Copyright 2013, American Chemical Society

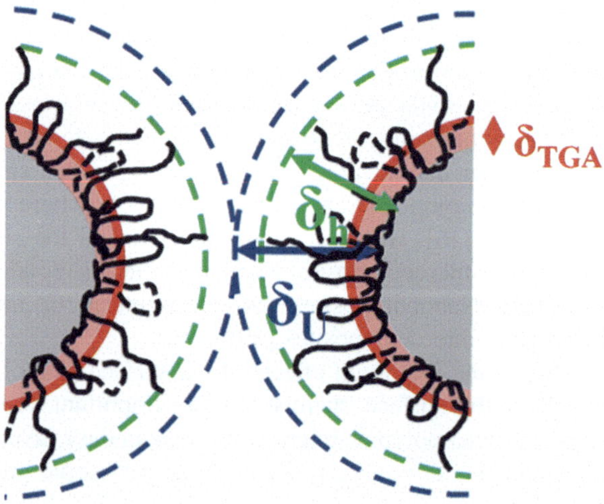

1.4.2 Challenges in Polymer Nanocomposites

The performance of polymer nanocomposites is affected by the characteristics of the constituents (including the aspect ratio, volume fraction, size, and specific surface area), as well as the interactions between phases. Even though great advancements have been made in polymer nanocomposites R&D over the last two decades, the dispersion of the nanoparticles in the polymer matrix remains a key challenge to their widespread application [35, 55]. Undoubtedly, to maximize the interfacial area in the polymer nanocomposites, the dispersion should be on the scale of the individual nanoparticles, otherwise aggregation results in a lower specific surface area, and microcomposites are ultimately formed. A large portion of this book is dedicated to reviewing methods for manipulating the interface, as well as the kinetic aspects of the dispersion of nanoparticles in polymer matrices (e.g., controlling extrusion parameters).

Another pressing matter is the control of the structural morphology of the polymer nanocomposite to achieve time-independent equilibrium. Often, further processing results in a different dynamic equilibrium, which makes it difficult to tune the emergent properties. A similarly complex challenge involves scaling-up preparation of the polymer nanocomposites to industrially viable quantities, especially for solvent-based systems. To secure economic and other societal impacts, sufficient volumes of nanomaterials need to be manufactured in order to be transformed into market-ready products. Most R&D labs around the world focus on processes with tightly controlled laboratory environments which are not necessarily appropriate for safe, reliable, effective, and affordable large-scale production. Funding to address these challenges is still required. This was suggested by Sargent [17] in a congressional research service report prepared as policy primer to members and committees of the US congress in 2016. Lastly, as with other new technologies, societal aspects need to be considered in parallel with development of the technology. Consumer perceptions regarding the risks of certain nanoparticles, whether proven or not, need to be addressed to ensure a smoother uptake of the technology.

In conclusion, the field of polymer nanocomposites has shown disparate results, successes as well as challenges. Much work remains to be done in fundamental research to achieve better control of the desired properties; for example, processing polymer composites to achieve the desired level of dispersion of the nanoparticles. This book investigates the effect of process variables on the structural evolution of polymer composites and the production of either micro- or nanocomposites, as well as the effect of the dispersion state on the final properties.

1.4.3 Scope of the Book

This book specifically discusses the processing–structure–property relationships of polymer nanocomposites at a fundamental level. Processing these nanocomposites usually requires special attention as the resultant structures on the micro- and nano-level are directly influenced by the polymer/nanoadditive chemistry and processing strategy, among others. The structure then affects the properties of the resultant composite materials. In Chap. 2, some common nanoparticles are introduced and we review some surface modification techniques employed to make the particles compatible with the host polymers. This has a direct implication on the dispersion of these particles and the resulting properties of the composites. Likewise, the important environmental and health and safety implications of the use of these particles are reviewed. As nanoclays are the most widely used nanofiller for preparation of polymer nanocomposites, a brief overview on these materials and their corresponding polymer nanocomposites is provided in Chap. 3. In Chap. 4, common characterization techniques are briefly presented. This is an important aspect in polymer nanocomposite technology as it helps researchers understand the dispersion state of the nanoparticles and elucidates the influence of the particles on the polymer matrices, e.g., studying boundary layer phenomena. Chapter 5 is the last part of the first volume and provides an overview of the effect of the melt-processing strategy on the structure and mechanical properties of polymer nanocomposites considering a clay-containing polymer nanocomposite as a model system. The first volume serves as an introduction to the second volume of the book, which focuses heavily on the processing technologies and strategies; we extensively discuss the processing–structure–property–performance relationships in a wide range of polymer nanocomposites, such as commodity polymers (Chap. 1), engineering polymers (Chap. 2), elastomers (Chap. 3), thermosets (Chap. 4), biopolymers (Chap. 5), polymer blends (Chap. 6), and electrospun polymers (Chap. 7). In particular, the important role played by nanoparticles in polymer blend structures is discussed.

Acknowledgements The authors would like to thank the Department of Science and Technology and the Council for Scientific and Industrial Research, South Africa, for financial support.

References

1. EU Commission. Commission Recommendation of 18 October 2011 on the definition of nanomaterial (2011/696/EU). Official Journal of the European Communities: Legis; 2011.
2. NSTC. The National Nanotechnology Initiative—Supplement to the President's 2017 Budget. USA: National Science and Technology Council; 2016.
3. The Japan Society for precision engineering. http://www.jspe.or.jp/wp_e/about_us_e/history/. Date Accessed. 06 June 2017.
4. National nanotechnology initiative. https://www.nano.gov. Date Accessed 6 June 2017.
5. LuxResearch. Nanotechnology update: corporations up their spending as revenues for nano-enabled products increase. Lux Research Inc.; 2013.
6. Roco MC, Mirkin CA, Hersam MC. Nanotechnology research directions for societal needs in 2020: retrospective and outlook. Springer Science & Business Media; 2011.
7. Dong H, Gao Y, Sinko PJ, Wu Z, Xu J, Jia L. The nanotechnology race between China and the United States. Nano Today. 2016;11:7–12.
8. Gao Y, Jin B, Shen W, Sinko PJ, Xie X, Zhang H, et al. China and the United States—global partners, competitors and collaborators in nanotechnology development. Nanomed Nanotechnol Biol Med. 2016;12:13–9.
9. Bai C. Progress of nanoscience and nanotechnology in China. J Nanopart Res. 2001;3:251–6.
10. Zhao D. China: a big player in a small world. ACS Publications; 2016.

11. BMMF. Action plan nanotechnology 2010. Federal Ministry of Education and Research; 2006.
12. BMMF. Action plan nanotechnology 2015. Federal Ministry of Education and Research; 2011.
13. BMMF. Action plan nanotechnology 2020. Federal Ministry of Education and Research; 2016.
14. RUSNANO Corporation. http://en.rusnano.com/about. Date Acceseed. 06 July 2017.
15. Bennett-Woods D. Nanotechnology: ethics and society. CRC Press; 2008.
16. Mnyusiwalla A, Daar AS, Singer PA. 'Mind the gap': science and ethics in nanotechnology. Nanotechnology. 2003;14:R9.
17. Sargent Jr JF. Nanotechnology: a policy primer. 2016.
18. Dunphy Guzman KA, Taylor MR, Banfield JF. Environmental risks of nanotechnology: National nanotechnology initiative funding, 2000–2004. ACS Publications; 2006.
19. Bacchini F. Is nanotechnology giving rise to new ethical problems? NanoEthics. 2013;7: 107–19.
20. Van de Poel I. How should we do nanoethics? A network approach for discerning ethical issues in nanotechnology. NanoEthics. 2008;2:25–38.
21. Allhoff F, Lin P, Moore D. What is nanotechnology and why does it matter?: from science to ethics. Wiley; 2009.
22. Hunt G, Mehta M. Nanotechnology:" Risk, Ethics and Law": Routledge; 2013.
23. Kaiser M, Kurath M, Maasen S, Rehmann-Sutter C. Governing future technologies: nanotechnology and the rise of an assessment regime. Springer Science & Business Media; 2009.
24. Grunwald A. Nanotechnology—a new field of ethical inquiry? Sci Eng Ethics. 2005;11: 187–201.
25. Bernard D. Ethics and nanomaterials industrial production. In: Metrology and standardization of nanotechnology: protocols and industrial innovations. 2017. pp. 485–504.
26. Coles D, Frewer L. Nanotechnology applied to European food production—a review of ethical and regulatory issues. Trends Food Sci Technol. 2013;34:32–43.
27. Bowman DM. More than a decade on: mapping today's regulatory and policy landscapes following the publication of nanoscience and nanotechnologies: opportunities and uncertainties. NanoEthics. 2017:1–18.
28. Schulte P, Roth G, Hodson L, Murashov V, Hoover M, Zumwalde R, et al. Taking stock of the occupational safety and health challenges of nanotechnology: 2000–2015. J Nanopart Res. 2016;18:1–21.
29. Maynard AD, Aitken RJ. 'Safe handling of nanotechnology' ten years on. Nat Nano. 2016;11:998–1000.
30. Ahn JJ, Kim Y, Corley EA, Scheufele DA. Laboratory safety and nanotechnology workers: an analysis of current guidelines in the USA. NanoEthics. 2016;10:5–23.
31. Donaldson K, Aitken R, Tran L, Stone V, Duffin R, Forrest G, et al. Carbon nanotubes: a review of their properties in relation to pulmonary toxicology and workplace safety. Toxicol Sci. 2006;92:5–22.
32. Sargent LM, Porter DW, Staska LM, Hubbs AF, Lowry DT, Battelli L, et al. Promotion of lung adenocarcinoma following inhalation exposure to multi-walled carbon nanotubes. Particle Fibre Toxicol. 2014;11:3.
33. Kojima Y, Usuki A, Kawasumi M, Okada A, Fukushima Y, Kurauchi T, et al. Mechanical properties of nylon 6-clay hybrid. J Mater Res. 1993;8:1185–9.
34. BCCResearch. NAN047F—2014 nanotechnology research review. Wellesley MA USA: BCC Research; 2014.
35. Kumar SK, Benicewicz BC, Vaia RA, Winey KI. 50th anniversary perspective: are polymer nanocomposites practical for applications? Macromolecules. 2017;50:714–31.
36. Giannelis EP. Polymer layered silicate nanocomposites. Adv Mater. 1996;8:29–35.
37. LeBaron PC, Wang Z, Pinnavaia TJ. Polymer-layered silicate nanocomposites: an overview. Appl Clay Sci. 1999;15:11–29.

38. Alexandre M, Dubois P. Polymer-layered silicate nanocomposites: preparation, properties and uses of a new class of materials. Mater Sci Eng, R. 2000;28:1–63.
39. Ray SS, Okamoto M. Polymer/layered silicate nanocomposites: a review from preparation to processing. Prog Polym Sci. 2003;28:1539–641.
40. Pavlidou S, Papaspyrides C. A review on polymer–layered silicate nanocomposites. Prog Polym Sci. 2008;33:1119–98.
41. Jordan J, Jacob KI, Tannenbaum R, Sharaf MA, Jasiuk I. Experimental trends in polymer nanocomposites—a review. Mater Sci Eng, A. 2005;393:1–11.
42. Okamoto M, Nam PH, Maiti P, Kotaka T, Nakayama T, Takada M, et al. Biaxial flow-induced alignment of silicate layers in polypropylene/clay nanocomposite foam. Nano Lett. 2001;1:503–5.
43. Baughman RH, Zakhidov AA, De Heer WA. Carbon nanotubes—the route toward applications. Science. 2002;297:787–92.
44. Moniruzzaman M, Winey KI. Polymer nanocomposites containing carbon nanotubes. Macromolecules. 2006;39:5194–205.
45. Thostenson ET, Ren Z, Chou T-W. Advances in the science and technology of carbon nanotubes and their composites: a review. Compos Sci Technol. 2001;61:1899–912.
46. Stankovich S, Dikin DA, Dommett GH, Kohlhaas KM, Zimney EJ, Stach EA, et al. Graphene-based composite materials. Nature. 2006;442:282–6.
47. Kim H, Abdala AA, Macosko CW. Graphene/polymer nanocomposites. Macromolecules. 2010;43:6515–30.
48. Verdejo R, Bernal MM, Romasanta LJ, Lopez-Manchado MA. Graphene filled polymer nanocomposites. J Mater Chem. 2011;21:3301–10.
49. Potts JR, Dreyer DR, Bielawski CW, Ruoff RS. Graphene-based polymer nanocomposites. Polymer. 2011;52:5–25.
50. Crosby AJ, Lee JY. Polymer nanocomposites: the "nano" effect on mechanical properties. Polym Rev. 2007;47:217–29.
51. Cheng S, Xie S-J, Carrillo J-MY, Carroll B, Martin H, Cao P-F, et al. Big effect of small nanoparticles: a shift in paradigm for polymer nanocomposites. ACS Nano. 2017;11:752–9.
52. Bauhofer W, Kovacs JZ. A review and analysis of electrical percolation in carbon nanotube polymer composites. Compos Sci Technol. 2009;69:1486–98.
53. Jouault N, Moll JF, Meng D, Windsor K, Ramcharan S, Kearney C, et al. Bound polymer layer in nanocomposites. ACS Macro Lett. 2013;2:371–4.
54. Hu H-W, Granick S, Schweizer KS. Static and dynamical structure of confined polymer films. J Non-Cryst Solids. 1994;172:721–8.
55. Müller K, Bugnicourt E, Latorre M, Jorda M, Echegoyen Sanz Y, Lagaron JM, et al. Review on the processing and properties of polymer nanocomposites and nanocoatings and their applications in the packaging, automotive and solar energy fields. Nanomaterials. 2017;7:74.

Chapter 2
Synthesis and Functionalization of Nanomaterials

Neeraj Kumar and Suprakas Sinha Ray

Abstract "Nanomaterial" and "nanotechnology" have become well-known terms, not only among scientists, engineers, fashion designers, and architects, but also the general public. Owing to their extraordinary and unexpected behavior, nanomaterials have gained tremendous attention in fields such as automotive, electronics, aerospace, healthcare, and biomedical, and have significant potential for many modern advanced technological applications. Nanomaterials have promised to make available products and systems smaller, better, lighter, and faster, which is achieving reality due to the rigorous efforts of scientists and engineers. In this scenario, several kinds of nanomaterials, various synthesis methods and advanced characterization techniques, and many computational models and theories to elucidate experimental results, are being developed by researchers. This chapter introduces state-of-the-art progress in the development of various synthesis strategies and functionalization approaches for producing a wide range of nanomaterials. We also discuss the properties of polymer nanocomposites considering some specific applications.

2.1 Introduction

Nanomaterials are a fundamental building block of material science and are gaining worldwide attention for various applications directly affecting our daily activities, in the medical science, electronics, energy, and environmental fields [1]. The distinctive tailorable size-dependent properties of these nanomaterials (over the scale

N. Kumar · S. Sinha Ray (✉)
DST-CSIR National Centre for Nanostructured Materials,
Council for Scientific and Industrial Research, Pretoria 0001, South Africa
e-mail: rsuprakas@csir.co.za

N. Kumar
e-mail: nkumar@csir.co.za

S. Sinha Ray
Department of Applied Chemistry, University of Johannesburg,
Doornfontein 2028, Johannesburg, South Africa
e-mail: ssinharay@uj.ac.za

S. Sinha Ray (ed.), *Processing of Polymer-based Nanocomposites*,
Springer Series in Materials Science 277,
https://doi.org/10.1007/978-3-319-97779-9_2

of 1–100 nm) make them attractive for various emerging applications [2]. The physical (structural, electronic, optical, and magnetic) and chemical (e.g., catalytic) properties of a material significantly improve when its size is reduced to the nanoscale regime due to quantum confinement effects [3]. Smaller materials show high surface area-to-volume ratios, which can be used to control the reaction rates (especially those dependent on surface area/defects) for many applications.

Nanotechnology is greatly contributing to sustainable solutions for critical challenges in the energy, environmental, and medical fields to sustain the ever-increasing global population by combining nanomaterials with engineering. Nanotechnology offers a spectrum of solutions: materials (powder/dispersed solution), polymer nanocomposites (PNCs), coatings, monoliths, prototypes, and devices [1]. Industries such as cosmetics, food, textile, pharmaceutical, construction, and automotive are already selling products with incorporated nanoparticles [4–6], benefitting from the last two decades of academic and industrial research in this area. The design and development of nanotechnologies is considered an interdisciplinary area of research which brings together theoretical and experimental researchers from various fields, such as catalysis, biology, polymer science, soft matter, colloid chemistry, physics, and electrical and chemical engineering. Interestingly, hybrid technologies, such as Li-batteries templated by viruses, have been developed due to the combined efforts of diverse research fields [7]. Similar efforts are expected to solve other existing manufacturing and microfabrication challenges. The demand for advanced nanomaterials has increased in commercial markets following demands for high quality, enhanced performance, and miniaturization of products, including solid sensors [8], electronic storage media [9], self-cleaning coatings [10, 11], healthcare products (e.g., sunscreens) [12], sponges for oil recovery [13], and efficient catalysts [13] and photocatalysts [14, 15]. In this scenario, the global market for nanomaterials is massive, billions of dollars, and expected to increase soon. Despite the hype of material science and nanotechnologies, only a few technologies have actually been integrated into end-user consumer products. This is mainly due to the high cost of the nanomaterials and challenges related to their large-scale production. Sometimes, these processes require elevated temperatures, or vacuum/inert gas atmospheres to facilitate the reactions. To form monodisperse nanoparticles (NPs), the reaction conditions must be very carefully monitored. Scaling-up laboratory recipes for nanomaterial synthesis to achieve mass production is complicated. It is challenging to produce materials with the same properties in each batch due to sensitive reaction conditions. This has resulted in a big obstacle to large-scale nanomaterial production and subsequently realizing their full potential in diverse fields. It is imperative to develop simple and cost-effective large-scale production processes and techniques to produce high-quality NPs.

Nanoparticle research began to receive attention from the early 1990s due to the discovery of the plasmonic and excitonic properties of metal and semiconductor NPs. Since then, different types of nanomaterials, including metal, polymeric, ceramic, and semiconductors NPs, carbon-based materials, and metal chalcogenides (MCs) have been explored. The fundamental properties of these materials have been primarily studied, along with their performance for anticipated new technologies.

Both nanotechnologies and fundamental research require nanomaterials with uniform morphology. This motivates researchers to design novel processes for nanomaterial synthesis, functionalization, purification, characterization, and post-synthesis modification. Nanomaterials can be synthesized using various processes, such as liquid-phase, solid-phase, and gas-phase methods. They can be produced directly as particles or dispersed in different hosts, such as liquids, glass, and gas (in the form of aerosol). Liquid-phase methods are the most popular due to their low cost, versatility, and simplicity. Such methods often use polar and nonpolar solvents and low temperature. For further advances, researchers studying NP synthesis still require a better understanding of nucleation and crystal growth processes, which is being achieved using advanced material characterization and modelling methods. This field is still in its infancy; many more novel processes and discoveries are expected.

Functionalization of nanomaterials is indispensable for realizing their full potential in e.g., the medical, engineering, and electronics fields, as well as for commercialization of nanoproducts. Several properties, such as the hydrophilicity, hydrophobicity, conductivity, and corrosion resistance of nanomaterials can be easily introduced by simple surface modification. Chemically functionalized nanomaterials (e.g., TiO_2, ZnO, Al_2O_3, Fe_2O_3, and ZrO_2) have been incorporated in organic/epoxy coatings to enhance performance and reduce toxicity [16, 17]. Inorganic NPs in the coatings also increase the scratch and adhesion resistance of the metal substrate. In clinical studies, targeted drug delivery and active cellular drug uptake have been achieved using multifunctional NPs with abetter efficacy. These multifunctional NPs can be obtained by conjugation with specific recognized chemical species [18]. Furthermore, homogeneous dispersion of nanomaterials in polymer solutions and organic solvents can allow exploitation of their full potential for a wide range of applications, such as ceramic, inks, paints, coatings, PNCs, and drug delivery. Uniform distribution of NPs can be achieved by simple modifications/functionalization using polymer/copolymers, biomolecules, surfactants, and other organic species. Proper dispersion of NPs in a polymer matrix can produce composites with improved mechanical, optical, electrical, and thermal properties. The development of materials with new and improved properties depends on the specific organization and dispersion of the NPs. Owing to their excellent properties, PNCs have attracted attention in various fields, including aerospace, packaging (e.g., cosmetics, electrical, and food), solar cells, construction, automobile, drug delivery, and environmental remediation. This chapter discusses state-of-the-art research into various nanomaterials, including synthesis and functionalization methods, properties, and applications in the context of polymer nanocomposites development.

2.2 Types of Nanomaterials

Nanomaterials can be broadly classified into two categories based on their chemistry: inorganic (non-carbon) and organic (hydrocarbon) nanomaterials. Metal-based NPs, metal oxide/hydroxide NPs, and transition metal chalcogenide

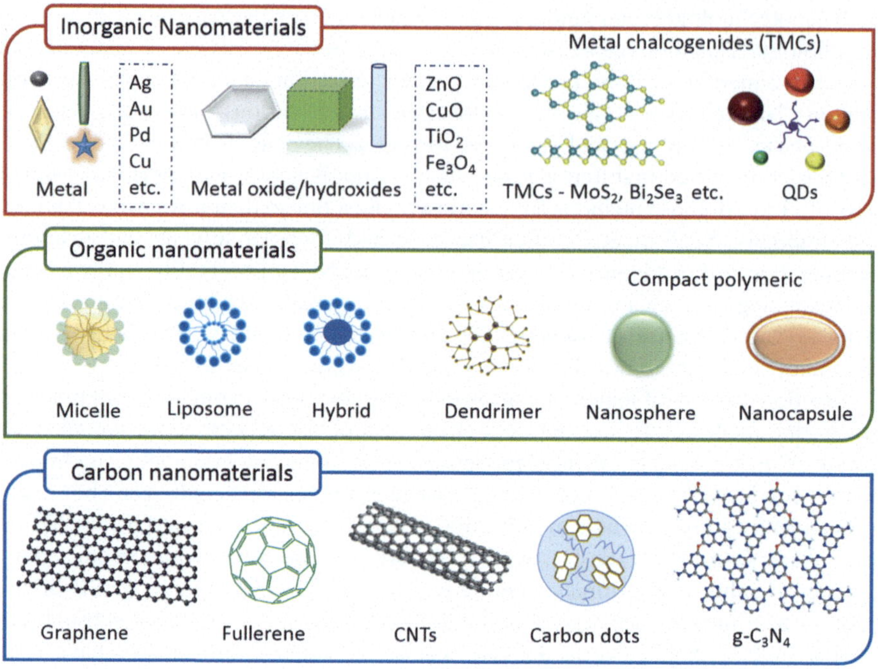

Fig. 2.1 Basic nanomaterials classification: **a** inorganic nanomaterials, **b** organic nanomaterials, **c** carbon-based materials

(TMC) NPs are inorganic nanomaterials. However, NPs based mainly on carbon and hydrogen are defined as organic nanomaterials, including compact polymers (nanospheres, nanocapsules), micelles, liposomes, dendrimers, hybrid NPs, and carbon-based nanomaterials. Due to the very large amount of studies and versatile applications, carbon-based nanomaterials (such as graphene, carbon nanotubes, and fullerenes) are considered as a separate class of nanomaterials with a wide range of morphologies. Figure 2.1 shows a schematic representation of this basic classification of nanomaterials.

2.2.1 Inorganic Nanomaterials

2.2.1.1 Metal NPs

Nanomaterials that consist of only one element are defined as metal NPs. They can exist as individual atoms or clusters of many atoms. In addition to existing in neutral forms (i.e., Ag(0), Au(0)), Au and Ag particles can exist in various nanocluster forms, such as Au_8, Au_{11}, Au_{13}, Au_{18}, Au_{25}, Au_{38}, Au_{55} and Ag_2 to Ag_8, and Ag_{25}, with characteristic electronic transitions [19]. Due to their excellent luminescent properties, such NPs are important for bio-labelling applications, producing luminescent patterns and fluorescence resonance energy transfer. Commonly synthesized NPs include Ag, Au, Cu, Pd, Pt, Re, Ru, Zn, Co, Al, Cd, Pb, Fe, and Ni; among them Fe and Ni are highly reactive and explosive. Further, metal NPs can also include the category of bimetallic NPs (i.e., Pt–Pd, Cu–Ni), which often exist as core–shell and alloy structures. Bimetallic NPs have better properties or efficiency than their single-metal NP counterparts [20]. Metal NPs are produced in the form of colloidal solutions or solid particles by simple techniques, such as bio-assisted synthesis, hydrothermal synthesis, microwave-assisted synthesis. They have shown interesting characteristics, such as localized surface plasmon resonance (LSPR), high reactivity, and broad absorption in the electromagnetic spectrum. Due to their advanced optical, optoelectrical, catalytic, anti-microbial/-cancer/-viral properties, metal NPs are highly interesting materials for numerous practical applications.

2.2.1.2 Metal Oxide NPs

Metal oxides are considered one of the most stable naturally occurring compounds. They are formed by reaction between electronegative oxygen and electropositive metal. They have polar surfaces due to presence of anionic oxygen and are insoluble in most organic solvents due to strong bonding between the metal and oxygen. The formation of metal oxides is the lowest free energy states for the metals in the oxidative nature of Earth and among other compounds of periodic table. Presently, they are widely used nanomaterials due to their high natural abundance, high chemical stability, tunable bandgap/band edge positions, and excellent thermal/electrical conductivity. Their applications include semiconductors, superconductors, and even insulators. With growing industrial interest, various types of versatile metal oxides, such as Al_2O_3, TiO_2, Fe_3O_4, Fe_2O_3, SiO_2, ZnO, and CeO_2 have been synthesized for application in the water purification, cosmetics, bio-medical, energy, and environmental remediation fields. These metal oxides can be easily modified by doping, resulting in hetero-structures and mixed oxides, to further meet the stringent demands of excellent properties and efficiency. (Layered) metal hydroxides are an interesting category of inorganic nanomaterials with flexible properties achieved by tailoring the structure and composition. These materials occur in two forms: (1) hydroxides with neutral layers, without intercalated molecules (i.e., β-$Ni(OH)_2$ and β-$Cu(OH)_2$); (2) hydroxides with a cationic layer and with intercalated molecules (i.e., α-$Ni(OH)_2$ and α-$Cu(OH)_2$) [21]. Due to their high surface area, high thermal stability, and excellent ion exchange capability, these materials have attracted attention for catalysis, supercapacitors, fuel cells, flame retardants, sensors, and pollutant removal.

2.2.1.3 Metal Chalcogenides (MCs)

Metal chalcogenides are a vast family of inorganic nanomaterials that contain at least one metal and one chalcogen anion (O^{2-}, S^{2-}, Se^{2-} and Te^{2-}). They are mostly restricted to sulfide, selenide, and telluride nanomaterials rather than oxide and polonium compounds due to the exceptional behavior of oxygen (gaseous and excellent non-metallic properties) and polonium (strong metallic properties). Since the 1960s, various transition metal (i.e., MoS_2) and post-transition metal chalcogenides (i.e., GaS) have been comprehensively explored due to their exciting properties (electronic and excitonic absorption) and quasi two-dimensional (quasi-2D) behavior [22, 23]. Recently, condensed matter physicists have shown tremendous interest in symmetry-protected topological states observed in bismuth selenide and tellurides [24].

Layered transition metal dichalcogenides (TMDs) are represented by the chemical formula MX_2 (where M is the transition metal and X is the chalcogen) and have van der Waals layer structures. Here, each individual monolayer consists of covalently attached X-M-X sandwich structures. Approximately forty types of layered TMD compounds can be formed from the elements in the periodic table [25] and they can occur in numerous polymorphs (metal coordination geometry) and polytype forms (stacking order). These layered TMDs can be categorized as follows: 2H (trigonal prismatic coordinated, 2 layer unit cell, hexagonal), 3R (trigonal prismatic coordinated, 3 layer unit cell, hexagonal), 1T (octahedral coordinated, 1 layer unit cell, trigonal), and distorted 1T′ [26]. 2H nanosheets are semiconducting, while other phases are metallic. MoS_2 is a typical TMD (group 6) nanomaterial well known for its catalytic, adsorption, and lubricant properties [27]. TiS_3 is an example of a transition metal trichalcogenide that has a distinctive quasi one-dimensional structure [28]. Bi_2Se_3 and In_2Se_3 have similar crystal structures consisting of quintuple layers [24]. Group 13 metals form metal monochalcogenides (i.e., GaS) in monolayers composed of four atomic layers [29]. SnS, SnS_2, and $SnSe_2$ chalcogenides are also semiconductors and have the same structure as TiS_2 [30]. These nanomaterials have bandgaps corresponding to the wide spectrum range from UV to near-IR wavelengths. The crystal structure of various classes of layered MCs and their bandgaps are shown in Fig. 2.2.

Due to their emerging interesting physical and technological properties, nanosheets of MCs have been broadly explored in recent years. Mono- and few-layer nanosheets of MoS_2 and its isoelectronic compounds (e.g., $MoSe_2$, WS_2, and WSe_2) show technologically important properties, such as piezoelectricity, optically accessible valley polarization, high exciton binding energy, direct bandgap photoluminescence (PL), and tunable many-body state [31, 35].

Due to their superior characteristics, MCs nanosheets are important building blocks for developing novel technologies for photonic and optoelectronic applications [36].

Quantum dots (QDs) are semiconductor nanocrystals (mostly 2–10 nm in diameter, comparable to the exciton Bohr radius) which have completely different electrical and optical properties than their bulk counterparts [37]. In principle,

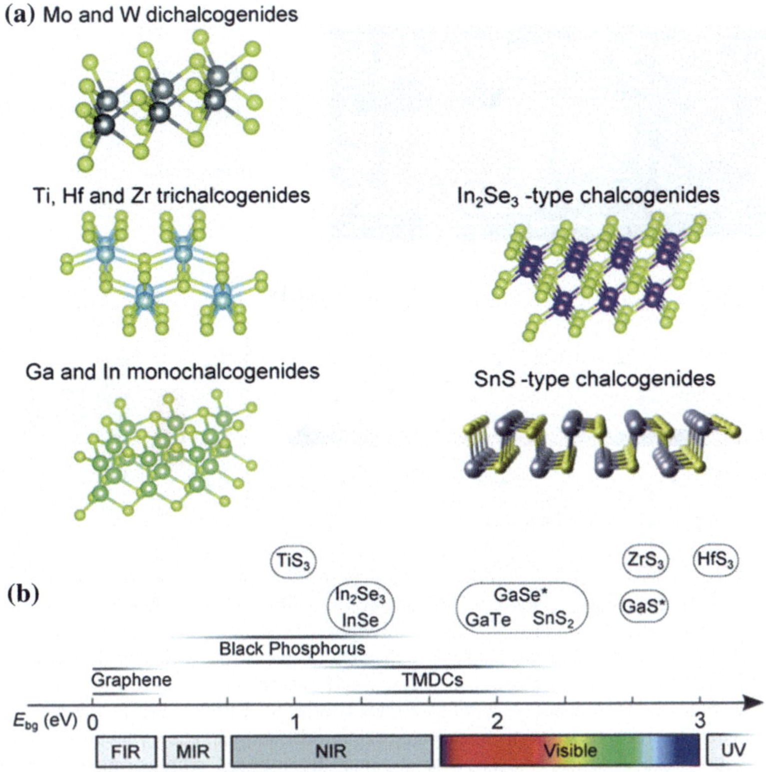

Fig. 2.2 **a** Representing the crystal structures of layered MCs nanosheets. Yellow colour atoms related to chalcogen. **b** Band gap of layered MCs nanosheets and other 2D nanosheets with respect to electromagnetic spectrum. Reprinted with permission from [31]. Copyright 2017, Springer Nature

the Bohr radius (r_B) is the average distance between the excited electron and hole that create an exciton in a bulk material. The Bohr model of the H atom can be expressed as:

$$r_B = \frac{\hbar^2 \varepsilon}{e^2}\left(\frac{1}{m_e} + \frac{1}{m_h}\right)$$

where $\hbar$, e, and ε represent the reduced Planck constant, charge of an electron, and dielectric constant of the semiconductor, respectively; m_h is the effective mass of the hole and m_e is the effective mass of an electron. The r_B values are not the same for each semiconductor as they depend on the effective masses of the hole and electron, and the dielectric constant of the material. It is noteworthy that a material with higher ε or smaller m_h and m_e, has a larger r_B. For example, r_B for ZnO is 2.3 nm [38], while it is 18 nm for PbS [39]. The motion of the electron and hole are

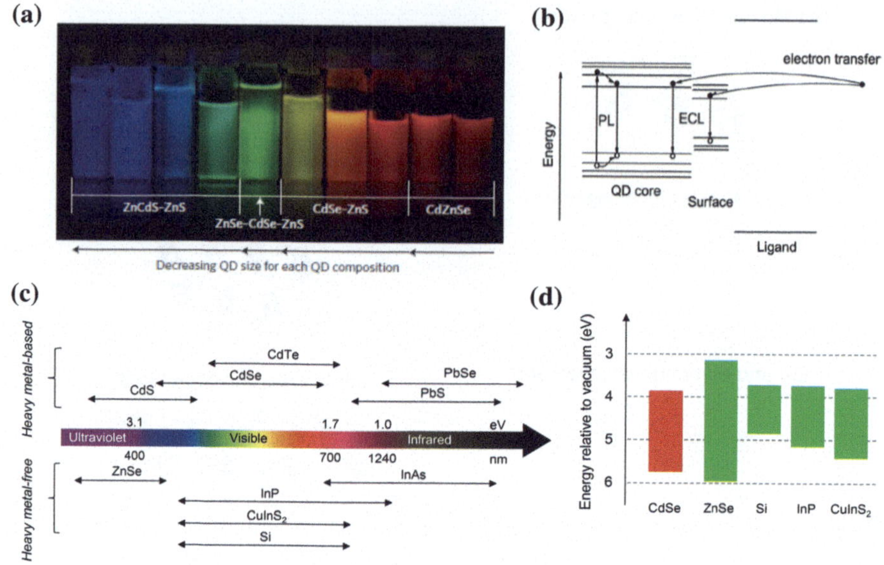

Fig. 2.3 **a** Size dependent photoluminescent color emission in different types of QDs. Reproduced with permission from [32]. Copyright 2013, Springer Nature. **b** Schematic representation of PL and ECL mechanism of Si QDs. Reproduced with permission from [33]. Copyright 2008, the American Chemical Society. **c** Represent heavy metal-based QDs (CdTe, CdS, CdSe, PbS and PbSe), heavy metal-free QDs (ZnSe, Si, $CuInS_2$, InP and InAs) and their respective emission from near-UV to far-infrared region. **d** Band-gap energy levels of bulk CdSe, Si, InP, $CuInS_2$ and ZnSe in respective to vacuum. Reproduced with permission from [34]. Copyright 2016, the American Chemical Society

confined within the spatial dimension of the QD (ultra-small particle) when the radius of the nanoparticle (R) approaches r_B of the semiconductor (i.e., $r_B \approx R$, $r_B > R$) [39, 40]. Due to quantum confinement, new properties (especially optical and electronic) develop in bulk semiconductors size reduced to QDs [41, 42]. Thus, QDs show size-dependent photoluminescence (PL) (Fig. 2.3a) electrochemiluminescence (ECL; electro-generated chemiluminescence), and absorbance [43]. Figure 2.3b) shows a schematic representation of the PL and ECL mechanisms in QDs. In fact, QDs can be easily modified to realize emission at wavelengths in the region of visible to far-infrared (450–1500 nm) by tailoring their morphology and composition [43]. Considering the requirements of the application, QDs can be synthesized as a colloidal solution in either aqueous or organic solvent. To date, diverse QDs, including classical metal chalcogenide QDs, Si QDs, carbon-based QDs, doped QDs, and core-shell QDs have been synthesized for various applications [44]. Metal chalcogenide QDs form the largest category of QDs. Chalcogenide QDs consist of hundreds of atoms of elements from group II–VI (e.g., ZnS, ZnSe, ZnTe, CdS, CdSe, CdTe), IV–VI (e.g., PbTe, PbSe, PbS), or I–III–VI_2 (e.g., $AgInS_2$, $CuInS_2$) [34]. Metal chalcogenide QDs exhibit different properties and emission spectra based on their bandgap and electronic states. Currently, QDs free from heavy metals such as Pb and Cd have gained huge attention from the research community in a desire for 'greener' QDs for sustainable processes [34]. Importantly, such QDs can easily replace heavy metal QDs in previously developed architectures as they have similar broad emission spectra, as well as band-edge energies comparable to CdSe QDs (Fig. 2.3c, d) [34]. The distinctive and diverse properties of QDs make them attractive for numerous applications, such as bioimaging, pharmacokinetics, cancer therapy, catalysis, sensing, supercapacitors, batteries, and optoelectronics [45, 46]. Very recently, QDs based on inorganic 2D materials have excited huge research interest due to their good biocompatibility, dispersibility, stability, improved optical properties, and tailorability [46].

2.2.1.4 Organic NPs

Liposomes, micelles, ferritin, dendrimers, hybrid molecules, and compact polymeric NPs are commonly categorized as organic NPs (Fig. 2.1). These particles are considered environmentally friendly as they are biodegradable and non-toxic. Materials with hollow cores (e.g., liposomes and micelles) are sensitive to electromagnetic (light) and thermal radiation (heat) energy [47]. Most of these NPs are ideal for drug delivery applications due to their high stability, biocompatibility, surface morphology, drug carrying capacity, and delivery efficiency. Liposomes are phospholipid vesicles and are composed only of lipidic compounds. The sizes of most liposomes are 50–100 nm, while unilamellar liposomes vary from 100 to 800 nm. The spherical structures of liposomes are formed by amphiphilic compounds. Importantly, they are biocompatible, almost completely biodegradable, versatile, non-toxic, non-immunogenic, and have good entrapment efficiency [48]. Micelle NPs are composed of amphiphilic molecules (viz. polymers or lipids) and have a size of 10–100 nm. They can provide on-demand hydrophilic and hydrophobic surfaces depending on the environment. For example, in aqueous medium, they only expose hydrophilic groups and conceal their hydrophobic parts inside the structure. They have high drug entrapment capacity, long circulatory, and high biostability [49]. Dendrimer NPs (<10 nm) are highly branched structures developed from one or more cores. The size of these NPs can be controlled by the number of generations that are grown over the cores. Dendrimer NPs are a nearly monodisperse polymer system synthesized by controlled polymerization with three structural parts: core, branch, and surface [50]. They are not a popular material for biomedical applications as incorporation and release of the drug is difficult and their synthesis is time-consuming. Compact polymeric NPs are formed as nanocapsules and nanospheres and consist only of synthetic or natural polymers. Their sizes vary in the range of 10–1000 nm. In this case, therapeutic agents can be trapped within the NPs as nanospheres or encapsulated within a polymeric shell as nanocapsules [51].

Thus, they allow complete protection of the drug and offer sustained localized drug delivery with decreased drug leakage. Various methods have been used to prepare polymeric NPs, including rapid expansion of a supercritical solution (RESS), rapid expansion of a supercritical solution into a liquid solvent (RESOLV), and surfactant-free emulsion polymerization [51]. This field is still quite immature and requires more focus on fundamental studies to develop materials suitable for application in various fields.

2.2.2 Carbon-Based NPs

The unique catenation properties of carbon enables it to form covalent bonds with other carbons in different hybridization states (sp, sp^2, and sp^3) to form a variety of structures of small molecules and long chains. Carbon is a versatile material and has diverse mechanical, chemical, optical, and electrical properties, which can be tailored by manipulating its structure and surface chemistry. The world of carbon-based materials is very rich, where the wide range of materials include graphite, diamond, amorphous carbons, fullerenes, graphene, nanotubes, nanohorns, nanocones, carbynes (linear carbon chains), and carbon onions (see Fig. 2.4)

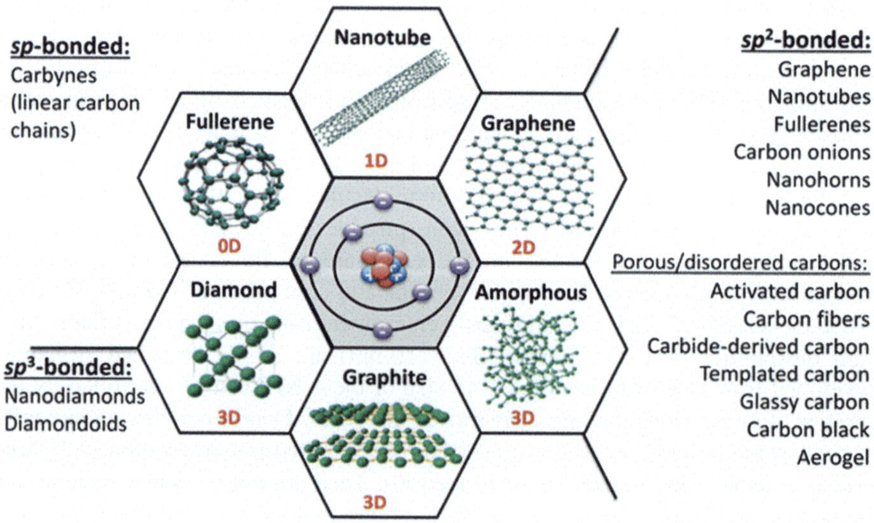

Fig. 2.4 The carbon nanomaterials categorized based on their dimensionality (i.e. zero-dimensional, 0D; one dimensional, 1D; two dimensional 2D; and three dimensional, 3D and bonding (i.e., hybridization states—sp, sp^2 and sp^3). Reproduced with permission from [54]. Copyright 2015, Cambridge University Press

[52, 53]. This diversity of nanostructures can make it difficult for engineers to select appropriate materials for specific applications. Materials such as activated carbon, carbon black, graphite, glassy carbon, diamond, and carbon fibers are commonly used in various industries. Other carbon materials, including nanotubes, nanodiamonds, nanocones, nanofibers, whiskers, nanorings, and nanohorns are being optimized for future technological applications [53].

The discovery of C_{60} in 1985 [55] and carbon nanotubes (CNTs) in 1991 [56] was accompanied by the development of many other carbon structures, including single-walled carbon nanohorns, bamboo-like nanotubes, and onion-like carbons spheres. In the quest for other carbon materials, the most recently extracted material is graphene, made from monolayers or a few layers of graphite in honeycomb-like hexagonal structures. This material was experimentally identified by Boehm et al. in 1962 [57]. Later, its extraordinary properties were studied by Geim and Novoselov in 2004 when they isolated and characterized single-layer graphene using the scotch tape method [58]. The graphene class consisting of monolayer or few-layer nanostructures was further extended to graphene oxide (GO), reduced graphene oxides (rGO), and graphene quantum dots (GQDs) by employing various solution-based synthesis routes with harsh oxidation conditions [52, 59]. Solution-based synthesis allows easy functionalization of these materials for applications from catalysis to biomedicine. Recently, there has been emerging interest in GQDs due to their unique optoelectronic properties. Likewise, graphite-like carbon nitride (g-C_3N_4), a polymer stacked into 2D layered structures with tri-s-triazine basic units, has been extensively studied for catalytic (especially photocatalyst), electronic, and energy applications [60].

The aforementioned carbon nano-allotropes can be grouped together as they are composed of hexagonally arranged sp^2 carbon atoms. They all have common properties including favorable optical properties, and high mechanical strength, conductivity, and chemical reactivity. Some variations in properties can be attributed to their different morphology. The dispersibility of carbon materials in organic solvents varies between materials. For example, the C_{60} nanostructure is highly soluble in organic solvent, while graphene can only be dispersed in selected solvents. Other carbon nanostructures form unstable dispersions in organic solvents. In parallel, carbon dots (C-dots) and nanodiamonds are being studied by researchers. C-dots contain sp^2 and sp^3 hybridized carbon atoms, while nanodiamonds consist only of tetrahedral sp^3 carbon atoms [54].

Recently, the synthesis and properties of carbon nano-allotropes (viz. fullerene, graphene, CNTs, and C-dots) have been extensively studied in view of numerous applications. Interesting materials or superstructures can be formed by carefully mixing different carbon nano-allotropes. Nanowires, 3D microspheres, carbon aerogels, monoliths, membranes, and thin films can be realized by appropriate combination of 0D, 1D, and 2D carbon nanostructures. Thus, carbon-based nanomaterials are a link between nanoscience and microarchitectures that are required for developing electronic-, optoelectronic-, photonic-, and biophotonic-based micro-devices. Research regarding chemical modification or functionalization to develop new reactive sites on carbon materials is enriching this field to fulfil the demands of new nanoproducts-based applications. Despite the excellent physico-chemical properties of CNTs and graphene, industrial application of these materials is very limited due to their high cost of production and the lack of suitable infrastructure. Thus, many companies have begun investing in the production of graphene and CNTs materials on the scale of tons for industrial applications.

2.3 Classification of Nanomaterials

Nanomaterials can be classified based on their dimensionality, composition, morphology, and uniformity or agglomeration state (Fig. 2.5). These topics will be discussed in the following sections.

2.3.1 Dimensionality

The dimensionality of nanostructures can be used to discriminate various types of nanomaterials. Firstly, Pokropivny and Skorokhod proposed the classification of

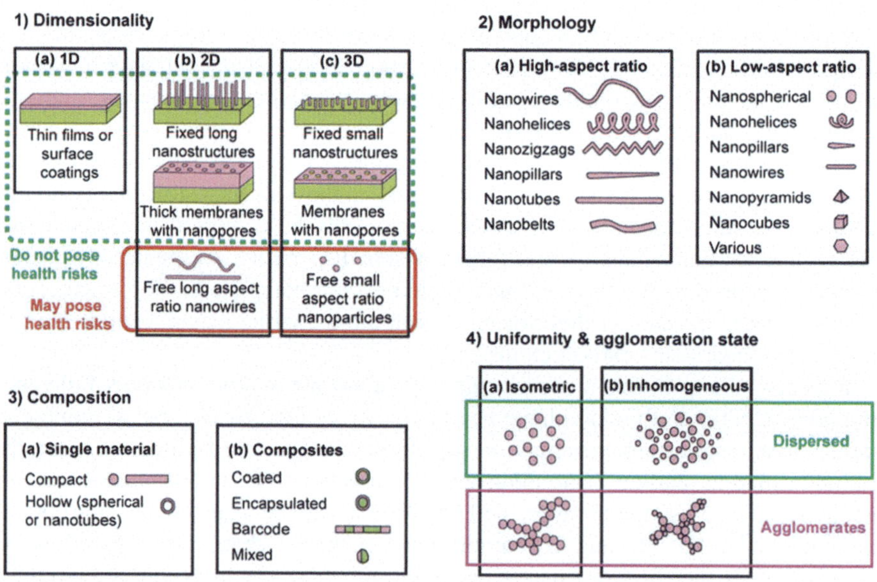

Fig. 2.5 Classification of nanomaterials based on dimensionality, composition, morphology, agglomeration state and uniformity. Reproduced with permission from [61]. Copyright 2017, the American Institute of Physics

nanomaterials based on 0D, 1D, 2D, and 3D nanostructural material behavior [62]. In the last decade, many studies have been conducted in the area of low-dimensional materials up to 3D micro/nanostructures. Numerous synthesis methods have been designed to fabricate such materials with uniform shapes and sizes. 0D nanomaterials include uniform nanoparticles, QDs, core-shell QDs/NPs, heterogeneous particle arrays, nanospheres, nano-onions, and nanolenses, which have been synthesized for solar cells, lasers, single electron transistors, and light emitting diodes (LEDs) applications [63]. 1D nanomaterials include nanoribbons, nanorods, nanospindles, nanowires, nanotubes, nanobelts, thin films or surface coatings, and hierarchical nanostructures. They play a highly significant role in electronic and optoelectronic nanodevices/systems, nanocomposites, and alternative energy resources as key components and interconnects between different phases. 2D nanostructural materials (NSMs) include nanosheets, nanoplates, nanoprisms, nanodisks, nanowalls, branched structures (dendrites), free long aspect ratio nanowires, nanoparticles attached to a substrate, junctions of continuous islands, and thin nanoporous films. In general, they have only one dimension in the nanoscale regime, for example, a thickness typically less than 5 nm and a lateral size in the range of 100 nm to few micrometers. The study of 2D nanomaterials has become the central materials research priority owing to their extraordinary characteristics that are completely different from bulk materials, making them suitable for a range of application. They also serve as basic building blocks for fabrication of nanodevices and super architectures. Some examples of ultrathin 2D nanosheets include TMDs, g-C_3N_4, layered double hydroxide (LDHs), layered metal oxides, hexagonal boron nitride (h-BN), graphene, GO, borophene, antimonene, black phosphorus (BP), silicene, metal-organic frameworks (MOFs), MXenes, covalent-organic framework (COFs), and perovskites [64]. Furthermore, 3D nanostructures include nanoflowers, nanocoils, nanopillars, nanoballs, dendritic-structures, small nanomaterials on substrates, and nanoporous membranes on a substrate. 3D nanostructures have the benefits of the nano-building blocks (2D and 1D materials) and concurrent advantages of the secondary micro-architectures. They exhibit a high surface area with 3D porosity, high physical absorption centers, and high structural stability. Figure 2.6 shows typical electron microscopy images of different types of 0D, 1D, 2D, and 3D nanostructured materials from the literature.

2.3.2 *Composition*

Considering the composition, nanomaterials can be classified as a single or composite material. Single materials consist of individual nanoparticles that can be hollow or compact. Composite materials consist of two or more materials.

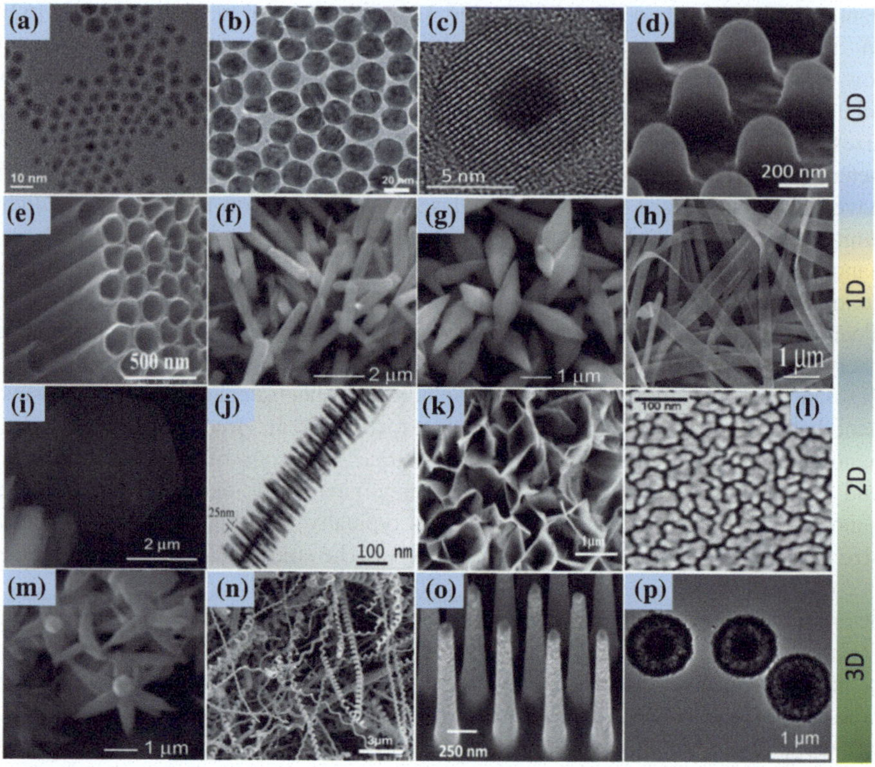

Fig. 2.6 Typical electron microscope images of different types of 0D, 1D, 2D and 3D NSMs. 0D NSMs: **a** QDs of Ag_2S. Reproduced with permission from [65]. Copyright 2011, the American Chemical Society. **b** Nanoparticles array of Au. Reproduced with permission from [66]. Copyright 2013, the Royal Society of Chemistry. **c** Core-shell QDs Au-CdTe. Reproduced with permission from [67]. Copyright 2010, the American Association for the Scientific Advancement. **d** Nanolenses. Reproduced with permission from [68]. Copyright 2013, the Royal Society of Chemistry. 1D NSMs: **e** nanotubes of TiO_2. Reproduced with permission from [69]. Copyright 2013, the Royal Society of Chemistry. **f** Nanorods of ZnO. Reproduced with permission from [15]. Copyright 2015, the Royal Society of Chemistry. **g** Nanospindles of ZnO. Reproduced with permission from [15]. Copyright 2015, the Royal Society of Chemistry. **h** Nanobelts of Ag. Reproduced with permission from [70]. Copyright 2014, the Royal Society of Chemistry. 2D NSMs: **i** nanosheets of ZnAl LDHs. Reproduced with permission from [71]. Copyright 2017, Elsevier Science Ltd. **j** Branched structures-dendrites of SiC. Reproduced with permission from [72]. Copyright 2010, the American Ceramic Society. **k** Nanowalls of ZnO. Reproduced with permission from [73]. Copyright 2017, Elsevier Science Ltd. **l** Nanoislands of Au. Reproduced with permission from [74]. Copyright 2009, the American Chemical Society. 3D NSMs: **m** nanoflowers of ZnO. Reproduced with permission from [15]. Copyright 2015, the Royal Society of Chemistry. **n** Nanocoils of C. Reproduced with permission from [75]. Copyright 2013, Elseviser Science Ltd. **o** Nanopillars of Si with SiO_2 layers. Reproduced with permission from [76]. Copyright 2014, Elsevier Science Ltd. **p** Yolk-shelled microspheres of V_2O_5. Reproduced with permission from [77]. Copyright 2013, Wiley-VCH

Coatings, barcodes, and encapsulated materials are examples of composite materials. They have attracted a huge amount of interest due to their enhanced mechanical, electrical, chemical, and thermal properties, as well as their good dimensional stability that have emerged from combining organic and inorganic hybrid materials.

2.3.3 Morphology

All developed morphologies can be classified into low- and high-aspect-ratio particles. Low-aspect-ratio nanostructures include nanospheres, nanocubes, nanopyramids, nanohelices, and NPs. The high-aspect-ratio nanostructures include nanozigzags, nanotubes, nanobelts, nanowires, nanosheets, and nanopillars. Such morphologies can be observed in some of the scanning electron microscopy (SEM)/ transmission electron microscopy (TEM) images presented in this chapter.

2.3.4 Agglomeration State and Uniformity

Nanomaterials can be further categorized based on the uniformity of their size distribution: homogenous (narrow particle size distribution) or inhomogeneous (unequal particle size). Depending on the agglomeration state, NPs can be grouped as dispersed or agglomerated particles. The degree of agglomeration depends on the surface charge and magnetism. The dispersion of NPs is governed by its surface morphology and type of functionalization, as well as the nature of the solvent. Surfactants/polymers are used to enhance the dispersibility of NPs. They can reduce surface tension of the water solvent and provide stable dispersion due to electrostatic or steric repulsion between surfactant molecules. The ultrasonication method is commonly used to improve the dispersion of NPs in different kinds of solvents.

2.4 Synthesis of Nanomaterials

Over the last two decades, various synthesis methods have been designed to achieve the aims of nanotechnology and nanofabrication. This field has witnessed significant growth in knowledge through the deep insights gained for different nanostructures and their performance for specific applications, as well the fabrication of nano-engineered materials for multi-purpose devices/systems. Further, new synthesis methods and fabrication technologies are advancing technological applications in the areas of nanomedicine, sensors, and electronics. The production of homogenous materials by cheap processes is a requirement for their widespread

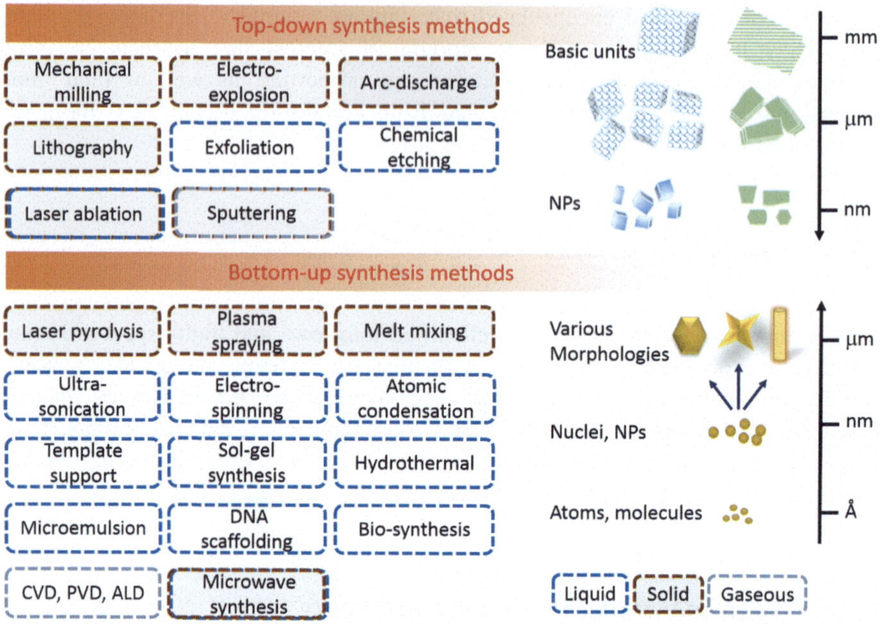

Fig. 2.7 Demonstration of various synthesis routes based on top-down and bottom-up synthesis methods. The rectangles in blue, brown and grey represent the liquid, solid and gaseous phase strategies

commercialization. Available synthesis methods are commonly divided into two main classes: (1) top-down and (2) bottom-up methods (Fig. 2.7).

Top-down synthesis is a destructive method where the reaction begins with starting materials that break down systematically into microparticles and are eventually converted into the desired NPs. Grinding/milling techniques, lithography, etching, sputtering, laser ablation, exfoliation, and electro-explosion are examples of top-down methods used for the synthesis of various nanomaterials [78]. For example, Zhou et al. reported top-down synthesis of photoactive Co_3O_4 NPs using laser fragmentation. The powerful laser produced homogenous Co_3O_4 NPs (size of 5.8 ± 1.1 nm) with a high number of oxygen vacancies [79]. Another study prepared magnetic NPs (size of ~20–50 nm) from high-purity natural iron oxide ore by a novel destructive top-down method using oleic acid [80]. Bottom-up synthesis methods involves assembling atoms and molecules to produce different kinds of NPs. Bottom-up synthesis approaches include: sol-gel, hydrothermal, microwave, and ultrasonication methods; self-assembly of monomer/polymer molecules; laser pyrolysis; plasma or flame spraying; electro-spinning; chemical or electrochemical precipitation; melt mixing; microemulsion processes; DNA scaffolding; template support synthesis; chemical vapor deposition (CVD); pulsed

Table 2.1 Comparison of various synthesis techniques based on different parameters including cost, yield, efficiency and controllability

Synthesis routes	Yield	Cost	Efficiency	Controllability	Remarks
Ball milling	High	Low	Low	No	Easy to introduce impurities, agglomerated particles, no morphology
Electrospinning	Medium	High	Medium	Low	Mainly nanoribbons >10 nm (aligned or random), required polymeric solution
Etching	Low	High	Low	Medium	Mainly nanoarrays
CVD, PVD, ALD	Low	High	Low	Low	Thin films and superlattice, high uniformity, toxic gases and precursors
Molecular beam epitaxy (MBE) Fast-quenching	High	Medium	Low	No	Ultra-fast cooling, speed needed, avoid phase transformation
Sputtering	Low	High	Low	Medium	Think films of uniform thickness, high temperatures and vacuums, need skilled operators
Deformation	High	Medium	Medium	No	Bulk materials should show good mechanical property
Wet chemical routes	Medium	Low	High	High	High controllability, close monitoring, several steps, diverse nanostructures

vapor deposition (PVD); atomic layer deposition (ALD); molecular beam epitaxy (MBE); and biological synthesis (using microorganisms and plant extracts). Wang et al. [81] demonstrated the synthesis of highly uniform spherical Bi NPs by both bottom-up and top-down methods. The obtained NPs had sizes of 100–500 nm. Bi was first melted in the top-down method and later emulsified in boiling diethylene glycol to generate monodispersed Bi NPs. In the bottom-up approach, Bi acetate was used as a starting material and diethylene glycol as solvent to produce Bi NPs [81].

Table 2.1 compares various synthesis techniques based on parameters such as cost, yield, efficiency, and controllability. In general, the aforementioned synthesis methods can be categorized into three groups: (1) liquid-phase methods,

(2) solid-phase methods, and (3) gaseous-phase methods (Fig. 2.7). Compared with other methods, liquid-based wet chemical methods are very facile methods for obtaining diverse nanomaterials with controlled shape, size, and morphology. Liquid-based wet chemical methods have the following advantages [82]:

- These processes allow appropriate selection of reaction parameters (e.g., types of solvents and/or surfactant, temperature, time) for the targeted product.
- The nucleation and growth of the NPs can be easily controlled by altering the thermodynamic and kinetic parameters of the reaction.
- Relatively simple processing conditions and low temperature.
- Low cost, as expensive instruments are not required. The cost can be further reduced by effectively designing reactions and carefully selecting chemicals.
- Liquid phase conditions result in good homogeneity and phase purity of the product.
- Diverse morphologies, such as nanorods, nanoneedles, nanosheets, nanoplates, nanoflowers, and hierarchical structures of the same materials can be achieved by simply varying the reaction conditions (e.g., salt precursors, pH, temperature, hydrolyzing agents, capping agent, solvent, and reaction methods).

As shown in the above discussion, wet chemical routes/liquid phase methods are efficient, and have a low cost, relatively high yield, and high controllability, allowing the fabrication of various nanostructures. Here, some common wet chemical routes are discussed in detail.

2.4.1 Hydrothermal Synthesis

In hydrothermal synthesis, the reaction medium is either water or a mixture of another solvent and water. Firstly, the reactants are mixed homogeneously in the reaction medium for a few minutes or hours, and then the mixture is placed in a Teflon-lined container. Subsequently, this container is placed in a stainless steel autoclave vessel which is tightly closed and transferred to a furnace to heat the material at a desired temperature. To develop autogenous pressure in the reactor by the pressure of vapor saturation, the temperature of the autoclave is usually above 100 °C. The in situ pressure developed within the autoclave is regulated by the reaction temperature, and the amount and type of added liquid and dissolved salts. This process has many advantages compared to conventional precipitation methods, including a short reaction time, fast kinetics, and suitability for large-scale production. In addition, materials with high crystallinity and phase purity can be prepared. Recently, the production of carbonaceous nanofibers was successfully achieved using a 12 L autoclave reactor while considering all parameters (time and temperature) that used for small volume reaction [83]. The diameter of the prepared nanofibers was precisely maintained even for this large volume. Thus, large-scale industrial production of nanomaterials could be achieved under suitable reaction

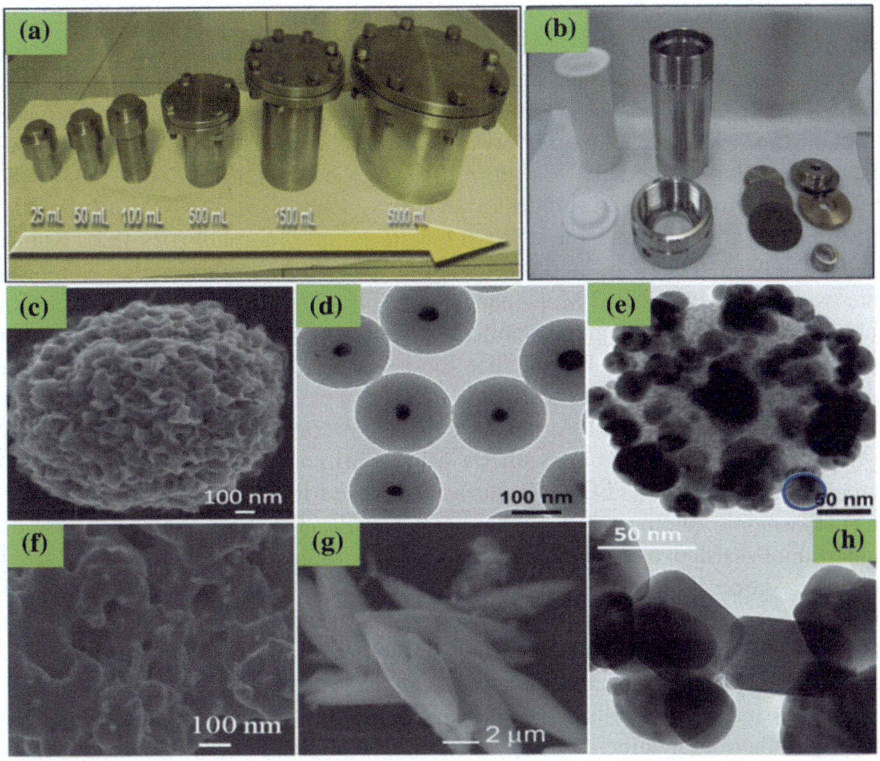

Fig. 2.8 **a** Photograph of Teflon-lined stainless steel autoclaves with different volumes; **b** shows all components of Teflon-lined stainless steel autoclave. Reproduced with permission of [82]. Copyright 2013, the Royal Chemical Society; Various nanomaterials synthesised by hydrothermal route: **c** hierarchical microspheres of PEGylated MoS_2 nanosheets. Reproduced with permission of [84]. Copyright 2017, Elsevier Science Ltd; **d** Ag@$MSiO_2$core-shell nanostructures [85], **e** Ag@SiO_2@Ag three core-shell nanostructures. Reproduced with permission of [85]. Copyright 2016, the Royal Society of Chemistry; **f** Ag/Ag_2Te nanocrystals. Reproduced with permission of [86]. Copyright 2017, the Royal Chemical Society; **g** micro-spindle structures of SrSe:Eu^{2+}. Reproduced with permission of [87]. Copyright 2015, Elsevier Science Ltd; **h** cubic HfO_2 nanoplates. Reproduced with permission of [88]. Copyright 2017, Springer Nature

conditions. Figure 2.8a, b show different sizes of Teflon-lined stainless steel hydrothermal reactors (25–5000 mL) and their components.

In this process, solvents (water and organic solvents) play a very important role as they readily dissolve reaction salts and introduce coordinating agents, such as ethylenediaminetetraacetate (EDTA), poly(vinyl pyrrolidone) (PVP), poly(ethylene glycol) (PEG), ethylenediamine, hydrazine hydrate, and ammonia, to control the crystal growth of NPs. The incorporation of small organic molecules as capping/stabilizing agents in this process can produce nanomaterials with various shapes and sizes. Zhao et al. illustrated the effect of organic ligands on the morphology evolution of copper sulfide NPs in an organic amine-assisted hydrothermal reaction at temperature range of 90–100 °C [89]. In this process, the monomer concentration and pH value play a very crucial role; the pH has a huge influence on the hydroxylation of inorganic and anion sources, while secondary anisotropic growth along the preferred axis is governed by the concentration of the monomer and reaction time. This technique is very useful for obtaining a wide range of nanomaterials, including metallic NPs (Pt, Zr, Au, Ag), metal oxides (ZnO, CuO, SnO_2, HfO_2, Fe_3O_4, In_2O_3), metal sulfides (MoS_2, SnS_2, ZnS, NiS), and NPs of $CoFe_2O_4$, SrSe, $FeWO_4$, Ag_2Te, and $La_{1-x}Sr_xCrO_3$ with unique morphologies [82]. Kumar et al. demonstrated a one-step hydrothermal route for synthesis of hierarchical microspheres of PEGylated MoS_2 nanosheets (Fig. 2.8c). PEG-400 plays a crucial role as a structure-directing and stabilizing agent for the growth of crystal facets of MoS_2 nanosheets. They showed that these nanosheets had cytotoxic behavior towards lung and breast cancer cells [84]. As seen in Fig. 2.8d, e, unique Ag@$MSiO_2$ core-shell and Ag@SiO_2@Ag three core-shell nanostructures were synthesized by hydrothermal synthesis using cetrimonium bromide (CTAB) and tetraethyl orthosilicate (TEOS) for surface-enhanced Raman scattering application [85]. Plasmonic photocatalyst nanocrystals (Ag/Ag_2Te) were synthesized by an ionic-liquid-assisted hydrothermal method using diphenyl ditelluride as a novel telluride source [86] (Fig. 2.8f). Intensely blue and red luminescent Eu(II, III)-doped SrSe microstructures were prepared using a hydrothermal route at comparatively low temperature using mixtures of water, hydrazine hydrate, and ethylene glycol [87]. Figure 2.8g shows the micro-spindle structures of blue phosphor SrSe: Eu^{2+}. Kumar et al. described the preparation of nanoplates of cubic HfO_2 by a post-hydrothermal treatment of previously prepared nanocrystals of cubic HfO_2 by a novel rapid microwave irradiation method (Fig. 2.8h). PEG-4000 was used as surface modifier to obtain cubic HfO_2 nanoplates, which were employed in breast cancer treatments [88].

2.4.2 Microemulsion Synthesis

Microemulsions are thermally stable, isotropic, optically transparent, and homogeneous dispersions. They consist of three main components: (i) a polar phase (i.e., H_2O), (ii) a non-polar phase (i.e., hydrocarbon liquid or oil), and (iii) a surfactant (i.e., CTAB). The surfactant plays an important role in decreasing the surface tension between the microemulsion and excess phase, and acting as a steric layer to prevent coalescence of the formed droplets by creating an interfacial barrier that isolates the organic and aqueous phases. Microemulsion systems are composed of monodispersed spherical droplets (diameter $\sim$8–600 nm) of water-in-oil and oil-in-water depending on the surfactant used. The reverse micellar (water-in-oil) system is most appropriate for the preparation of NPs [90]. In reverse micellar systems, polar heads of surfactant reside inside spherical droplets and create an aqueous core, whereas non-polar hydrophobic hydrocarbon tails point outwards (see Fig. 2.9a).

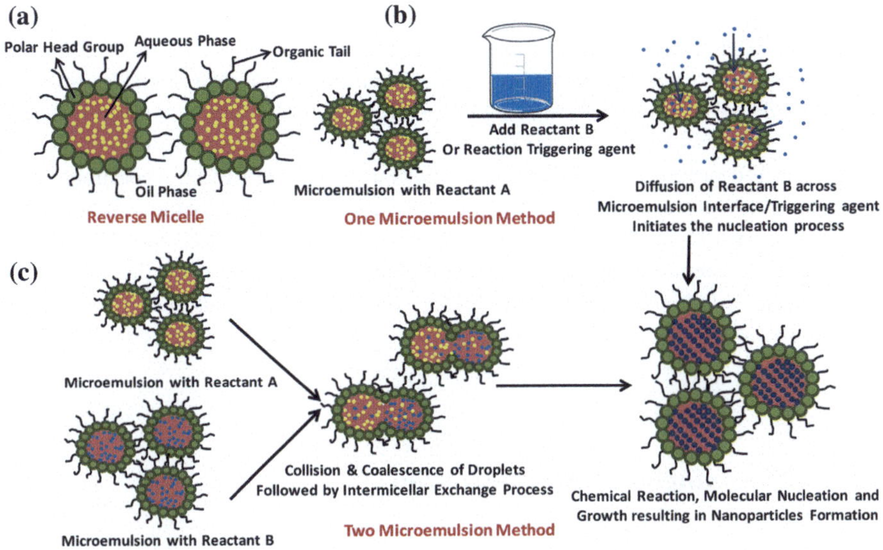

Fig. 2.9 Demonstrates **a** the reverse micelle system, **b** numerous step of one microemulsion NPs process, **c** reaction steps involved in the two microemulsion method. Reproduced with permission from [91]. Copyright 2015, the Royal Society of Chemistry

Nanomaterials can be synthesized in two ways using the microemulsion method, termed the one microemulsion and two microemulsion methods. In the one microemulsion method, a microemulsion of a particular reactant is prepared and nucleation can be initiated either by adding an energetic trigger agent (high power radiation/reducing agent/oxidizing agent) or second reactant [91]. This method is diffusion controlled as the triggering agent/second reactant has to diffuse to the interfacial walls of the microemulsion that carried the first reactant to generate NPs (Fig. 2.9b). In the case of the two microemulsion method, microemulsions of two separate reactants are mixed in different ratios to form micro-reactors by intermicellar collisions and coalescence of spherical droplets (Fig. 2.9c). The vigorous stirring or addition of any external chemicals (such as precipitating agents, reducing agents, and oxidizing agents) initiates fast nucleation and crystal growth of NPs. Microemulsion processes have been employed for synthesis of inorganic and organic nanomaterials. Inorganic nanomaterials, such as metal NPs (Pd, Pt, and Au), metal salts ($SrCO_3$, and $BaCO_3$), magnetic NPs ($ZnFe_2O_4$, (Ni, Zn)Fe_2O_4, and $BaFe_{12}O_9$), metal oxides (TiO_2, ZnO, ZrO_2, SiO_2, and Fe_2O_3), metal sulfides (CuS, CdSe, and PbS), and composites (CdS–TiO_2 and CdS–SnO_2) have been prepared using this technique [91]. A diverse class of nanomaterials with core–shell structures can also be developed using microemulsion techniques. For example, bimetallic core–shell Pd–Ag and Pd–Au NPs with sizes in the range of 10–30 nm have been synthesized using the non-ionic surfactant TX-100 by a seed-mediated method [92]. Moreover, the microemulsion method can be used to control the size and morphologies of NPs by varying parameters such as the concentration of surfactant, type of surfactant (anionic, cationic, non-ionic, and zwitterionic), type of continuous phase, and amount of precursors. Disadvantages of this process are the requirements for excessive washing and further chemical treatments to stabilize obtained NPs, which are generally agglomerated. Solla-Gullan et al. demonstrated scalable synthesis of cubic Pt NPs using water-in-oil microemulsion by reducing H_2PtCl_6 in the presence of sodium borohydride [93]. Yildirim et al. demonstrated the synthesis of uniform ZnO NPs by microemulsion using sodium bis (2-ethylhexyl)sulfosuccinate as surfactant. Firstly, agglomerated zinc glycerolate particles were prepared using the microemulsion process; then the samples were subsequently calcined to obtain ZnO NPs [94]. Conducting polymer NPs (such as polyaniline, polythiophene, and polypyrrole) were also synthesized using this process [91].

2.4.3 Biosynthesis of Nanomaterials

Biosynthesis is considered a green, environmentally benign, non-toxic, low-cost, and efficient approach to synthesize various NPs. Biological systems such as yeast, bacteria, viruses, fungi, plant extracts, and actinomycetes have been used to synthesize various nanomaterials, including metal NPs, metal oxides, metal sulfides, and carbon-based materials [95]. Nanomaterials obtained using this process are highly biocompatible, which is a prime requirement for medical use of NPs. Currently, extensive research efforts are being undertaken to synthesize NPs using plant extracts, biomolecule templates, and microorganisms [91]. Bio-assisted synthesis of NPs using plant biomass or plant extracts is comparatively faster than other biosynthesis methods and has predominantly been used to synthesize metallic NPs, bimetallic alloys, and metal oxides NPs. Yun et al. identified many plant biometabolites based on their reducing and capping abilities in the preparation of NPs [95]. Plant extracts containing chemicals such as amino acids, citric acid, tannic acid, tartaric acid, terpenoids, polyols, dehydrogenase, flavonoids, proteins, and reducing sugars have been used as reducing agents. Shankar et al. demonstrated the use of *Azadirachta indica* leaf extract to synthesize Ag, Au, and core–shell Au–Ag NPs [96]. Seraphin et al. used an aloe vera plant extract to prepare In_2O_3 NPs (5–50 nm) [97]. Extracts isolated from different parts of diverse plants have been employed to prepare Pd and Pt NPs [98, 99]. Figure 2.10 illustrates the mechanism behind the bio-assisted synthesis of metal NPs using plant extracts.

Tremendous research efforts have been undertaken to produce various NPs (e.g., Au, Ag, Pd, SiO_2, TiO_2, ZrO_2, and CdS) using microorganisms (prokaryotic bacteria, algae, yeast, fungi, and actinomycetes) as bio-reactors [91]. Microorganisms convert the target ions into metal NPs using enzymes produced during cellular activities. Boon et al. reported non-enzymatic synthesis of Ag NPs using lactic acid bacteria (*Lactobacillus* spp.) both as capping and reducing agents; interestingly, marine yeast (R. diobovatum) was used for biosynthesis of PbS NPs [100]. Similarly, other authors reported the production of Fe_3O_4, SiO_2, and TiO_2 using

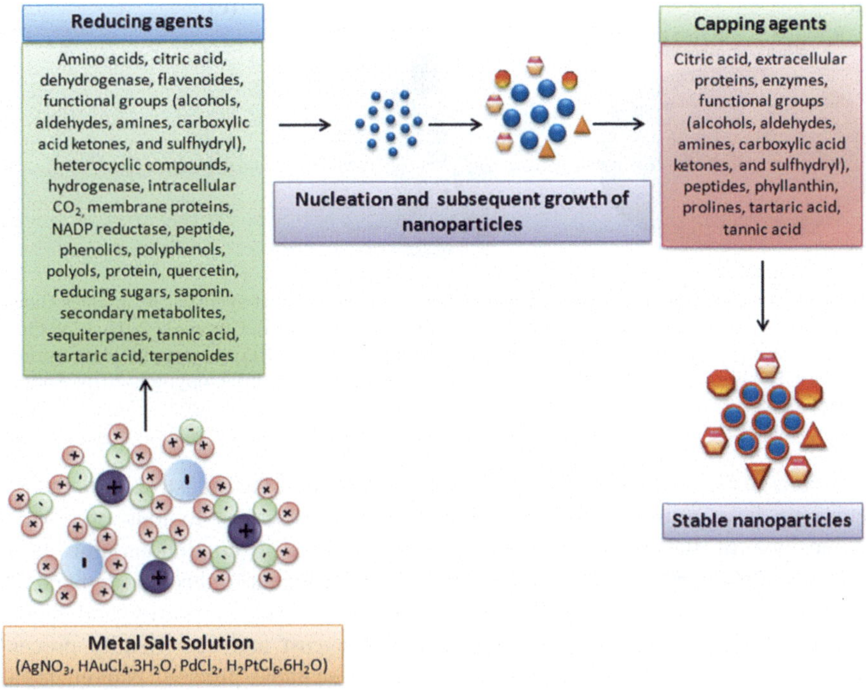

Fig. 2.10 Schematic illustrates the mechanism of biosynthesis of metal NPs using plant extracts. Reproduced with permission from [95]. Copyright 2013, the American Chemical Society

different types of fungi [101, 102]. Biomolecules including diatoms, membranes, viruses, and nucleic acids (RNA, DNA) have been used as templates for the production of various NPs. For example, Kundu et al. demonstrated the synthesis of ZnO nanostructures with diverse morphologies, including wires, flowers, and flakes using DNA as biotemplates and microwave heating [103]. Compared to other synthesis techniques, biosynthesis is quite slow, requiring many hours or even days. This process cannot yet achieve nanoparticles with the desired shape, size, and morphology. Moreover, biosynthesizes NP may decompose after a period of time.

2.5 Large-Scale Production of Nanomaterials

Over the past decade, nanomaterials have played a significant role in technological developments in engineering, electronics, medicine, and many other applications. Although there are myriad applications of these nanomaterials in literature, nanomaterial-based products are still limited in the market. This gap occurs due to the lack of efficient and low-cost manufacturing processes and the reluctance of industries to adopt and invest in new infrastructures. Large-scale production of

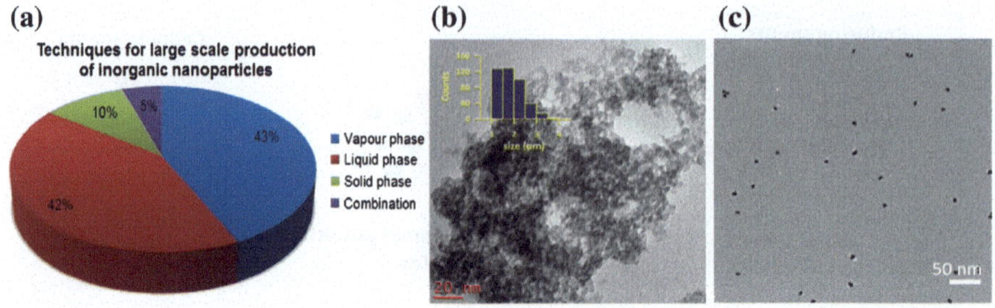

Fig. 2.11 **a** Large-scale production of nanomaterials using different synthesis techniques (liquid, solid, gaseous, and combination). Reproduced with permission [104]. Copyright 2016, EDP Open; **b** TEM image of ZnS:Ag (%) NPs. Reproduced with permission [105]. Copyright 2016, Springer Nature; **c** TEM image of Pt NPs. Reproduced with permission [106]. Copyright 2016, Springer Nature

nanomaterials should meet the requirements of tests that the industries undertake to determine is a process is suitable. This requires kilograms to tons of powder, suspensions of flakes, or slurries, or many pieces of continuous thin films. Nanomaterials need to be produced based on stringent industrial demands to achieve: (i) the desired properties and morphology; (ii) the purity and quality sufficient for the targeted application; (iii) a low-cost and eco-friendly process; (iv) a stable and controllable process; and (v) a scalable process. Nanomaterials have already entered the value chain but are not commonly seen in manufactured products. Carbon-based nanomaterials, metal oxides NPs, and nanoclays are synthesized globally on a large scale [104]. Many industries have started using nanomaterials in their production line, but in very small quantities due to their high cost. Globally, large-scale production of inorganic nanomaterials mainly uses liquid-phase and vapor-phase techniques [104] (Fig. 2.11a).

Moon et al. developed a pilot-scale process for microbial-mediated synthesis of ZnS and doped ZnS NPs ($\sim$2 nm) using a 900 L pilot plant reactor [105] (Fig. 2.11b). In many studies, continuous hydrothermal flow synthesis was used to synthesize various inorganic NPs. Riche et al. demonstrated the synthesis of Pt NPs by 3D-printed droplet microfluidic generators using ionic liquid as solvent (Fig. 2.11c). Yield of the product significantly increased using microfluidic reactor processes compared to batch synthesis [106]. The market demands for CNTs, graphene, GO, QDs, and their hybrids are continuously increasing. Graphene is produced industrially using mainly CVD, exfoliation of graphite, and exfoliation or reduction of graphite oxides. Industrial CNTs are generally synthesized in one of three different ways: using CVD, laser ablation, and electric-arc discharge. In CVD and laser ablation processes, the dimensions of the CNTs can easily be controlled. The laser ablation method is efficient and provide maximum yield. MWCNTs (Multi-walled CNTs) have already been produced in a bulk quantity by many companies such as Nanocyl SA, Shenzhen Nanotech Port Co., Ad-Nano technologies, Reinste Nano Ventures and Cnano Technology Ltd [107]. At this stage, the highest production (400 tonsCNT/year) of MWCNTs is carrying out by Nanocyl SA using mobile-bed technology with continuous rotating reactor system [108]. Moreover, continuous flow production systems using microfluidic channels and colloidal synthesis are most commonly used for mass production of QDs.

2.6 Functionalization of Nanomaterials

Control of the dispersion of NPs in different matrices and the interactions between NPs and other molecules is necessary to explore their full potential for biomedical, nanocomposite, sensing, imaging, energy, electronic, and other applications. This can be achieved by proper surface functionalization and modification of the NPs, which enhance their interaction with surrounding environments. Functionalization of NPs consequently influences the colloidal stability, dispersion, and controlled assembly of NPs. For example, for biomedical applications, surface functionalization with biomolecules can provide stealth properties, high cellular internalization ability, efficient intracellular delivery, and overall high biocompatibility [18]. Surface modification can also provide selectivity or specific recognition in biosensing and targeted delivery in biomedicine. By tailoring the surface of NPs, properties such as hydrophobicity, hydrophilicity, conductivity, and corrosivity can be readily introduced by adjusting the surface environments. Hydrophobic NPs can be produced by modification during synthesis or post treatments using hydrophobic molecules, such as oleylamine (OAM), oleic acid (OA), dodecanthiol (DDT), triphenylphosphine (TPP), trioctylphosphine oxide (TOPO), and tetraoctylammonium bromide (TOAB) (Fig. 2.12a). Likewise, hydrophilic NPs can be produced by modification using hydrophilic molecules such as polyethylene glycol (PEG), aminenated PEG, mercaptosuccinic acid (MSA), bis-sulfonated triphenylphosphine

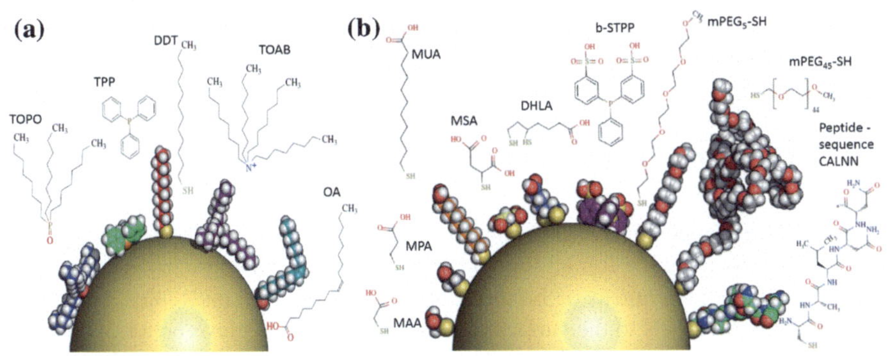

Fig. 2.12 Surface modification of NPs using hydrophobic ligand molecules (**a**) and hydrophilic ligand molecules (**b**). Schematically spatial conformation of ligands is obtained from their space-filling models and chemical structures. Reproduced with permission from [109]. Copyright 2010, the Royal Society of London

(b-STPP), mercaptoacetic acid (MAA), mercaptoundecanoic acid (MUA), mercaptopropionic acid (MPA), dihydrolipoic acid (DHLA), mercaptosuccinic acid (MSA), short peptide of the sequence CALNN, $mPEG_5$-SH, and $mPEG_{45}$-SH [109] (Fig. 2.12b).

Although surface engineering of nanomaterials is not an easy task, researchers have an efficient toolbox for adding surface functionality through in situ synthesis and post-synthesis modifications. NPs have been functionalized using various synthesis strategies with a diverse range of ligands, such as surfactants, polymers, small molecules, polymer, dendrimers, biomolecules, and inorganic materials. Considerable attention and research efforts have been devoted to the study of nanomaterial interface modification and functionalization. Various developed synthesis strategies for functionalization of NPs include:

- The use of chemisorption (e.g., thiol groups attached on the surface of NPs) and physisorption (milling and/or mixing).
- Electrostatic interactions: negatively charged NPs attract positively charged small molecules or vice versa.
- Covalent interactions: using conjugation chemistry by exploiting functional groups of both NPs and different molecules.
- Non-covalent or supramolecular affinity: based on receptor-ligand systems and weak interactions.
- Intrinsic surface engineering: heteroatom incorporation and defect engineering.

The aforementioned synthesis strategies can be effectively used to functionalize nanomaterials. Figure 2.13 shows a schematic of the types of modifications and subsequent adjustment of numerous surface properties, such as the topology, surface charge, surface energy, reactivity, and bioactivity; the surface properties control the overall behavior and efficacy of the materials. Thus, surface functionalization can provide the properties required for a particular application while retaining bulk material properties. A current trend is multiple modifications using various modifiers on the same substrate. In the next sections, molecular functionalization (covalent and non-covalent) and intrinsic surface engineering (heteroatom incorporation and defects engineering) are discussed in detail.

2.6.1 Covalent Functionalization

Covalent functionalization allows the attachment of different chemical species to the nanomaterials via the formation of covalent bonds. This approach is commonly used to enhance attachment of biomolecules, small organic molecules, polymers, and inorganic materials to the surface of nanomaterials to achieve better dispersion, high colloidal stability, and versatile properties. Covalent functionalization is attractive as it is regularly employed for modification of common nanomaterials, including metal oxides NPs, 2D nanomaterials (e.g., GO, CNTs, MoS_2, and h-BN), and nanoclays to optimize materials for a range of applications, including

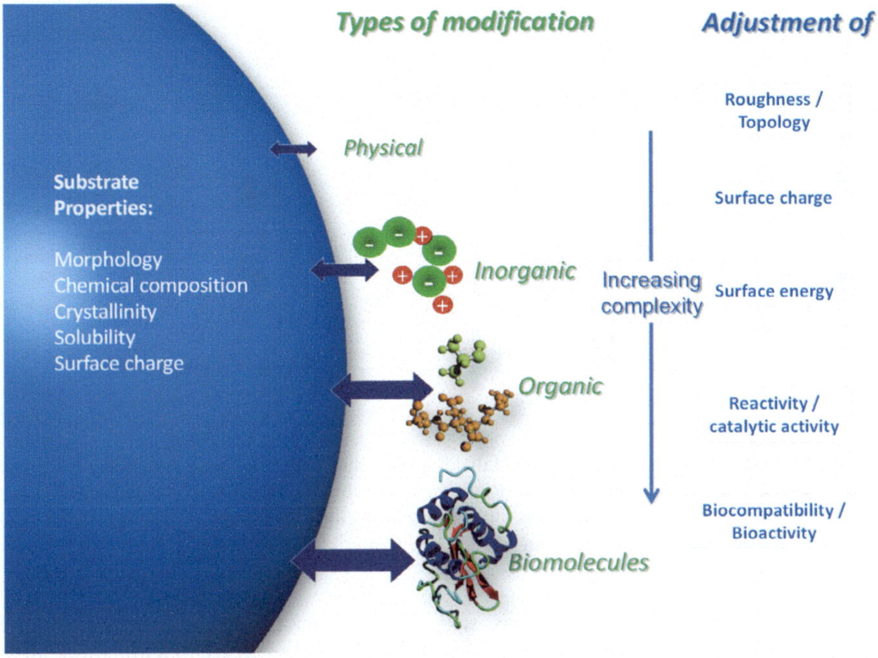

Fig. 2.13 Illustrates the substrate properties, types of modifications and consequently adjustment of numerous properties. Reproduced with permission from [110]. Copyright 2013, Elsevier Science Ltd

biosensors, bioimaging, environmental remediation, packaging, cosmetics, agriculture, and catalysis.

Various functionalization protocols have been developed using heterobifunctional cross-linker molecules. The most common methods for covalent functionalization are based on organofunctional alkoxysilanes ((3-aminopropyl)-triethoxysilane (APTES), (3-glycidoxypropyl)-dimethyl-ethoxysilane (GPMES), (3-mercaptopropyl)-trimethoxysilane (MPTMS), etc.), glutaraldehyde (GA), *N*-hydroxysuccinimide (NHS) and 1-ethyl-3-(3-dimethylaminopropyl)carbodiimide (EDC) chemistry [110]. These protocols can be applied for functionalization of oligonucleotides, proteins, peptides, and several NPs. The basic reactions required for silanization, GA, NHS, and EDC protocols are schematically illustrated in Fig. 2.14. Figure 2.14a shows the immobilization of biomolecules using silanization and the GA protocol, which involves: (i) oxide surface silanization using APTES; (ii) reaction of amine-terminated substrate with GA; (iii) attaching biomolecules via covalent bonds; and (iv) stable imine bond formation between the biomolecule and substrate [111]. Figure 2.14b shows the immobilization of biomolecules using the EDC/NHS method: (i) EDC reacts with the carboxyl group of the biomolecule and forms unstable intermediate *O*-acylisourea; (ii) NHS reacts with the intermediate *O*-acylisourea and forms stable amine-reactive NHS ester;

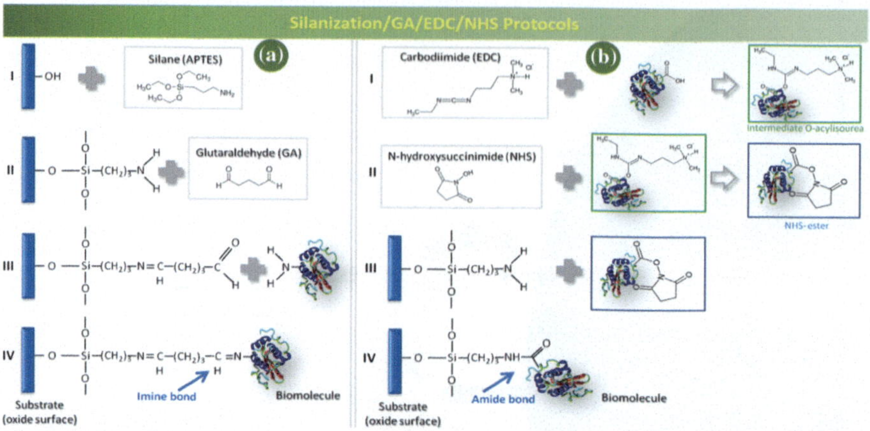

Fig. 2.14 Illustrates the covalent functionalization of oxide substrate with biomolecules using silanization/GA/EDC/NHS protocols. **a** Represents the immobilization of biomolecule using silanization and GA approach. **b** Shows the immobilization of biomolecule using EDC/NHS approach. Reproduced with permission from [110]. Copyright 2013, Elsevier Science Ltd

(iii) amine-terminated surface reacts with NHS ester; (iv) formation of a stable amide bond, and ultimately, biomolecule immobilization [110]. Using these protocols, enzymes, proteins, DNA, RNA, inorganic complexes, and small ligands can be irreversibly functionalized on the activated surfaces via covalent interactions between amine-terminated/aldehyde substrates and carboxyl/amine/hydroxyl groups of guest molecules. Several nanomaterials, such as layered double hydroxides, metal oxide NPs, GO, oxidized-CNTs, hydroxyl-functionalized boron nitride nanosheets (OH-BNNSs), and hydroxyl/carboxyl terminated TMCs can be modified using these protocols.

Many synthesis strategies have produced 2D nanomaterials by targeting their point defects, edges and basal functionalities, and π-electron clouds (converting sp^2 carbon to sp^3 carbon) [112–114]. Conversion of sp^2 carbon to sp^3 carbon in most carbon-based 2D materials results in defects or distortions on the surface and subsequent changes to the electronic properties. Sainsbury et al. demonstrated covalent functionalization of exfoliated boron nitride nanosheets (BNNSs) using two steps and its polymer nanocomposite [115]. Firstly, alkoxy groups were grafted onto BNNSs in the presence of di-tert-butyl peroxide; secondly, hydrolytic defunctionalization of the group was performed to synthesize OH-BNNSs (Fig. 2.15a). In another report, 3-aminopropyltrimethoxysilane (APTMS) was covalently functionalized on GO using a silanization protocol [113]. Later, APTMS-coated graphene was functionalized by an oxo-vanadium Schiff base and applied for catalytic oxidation of alcohols (Fig. 2.15b). He et al. reported covalent functionalization of single- and multi-walled CNTs with heteroarene iodonium salts and substituted arene [116]. The functionalization of phenyl groups carrying electron withdrawing groups on CNTs was illustrated by reactions with

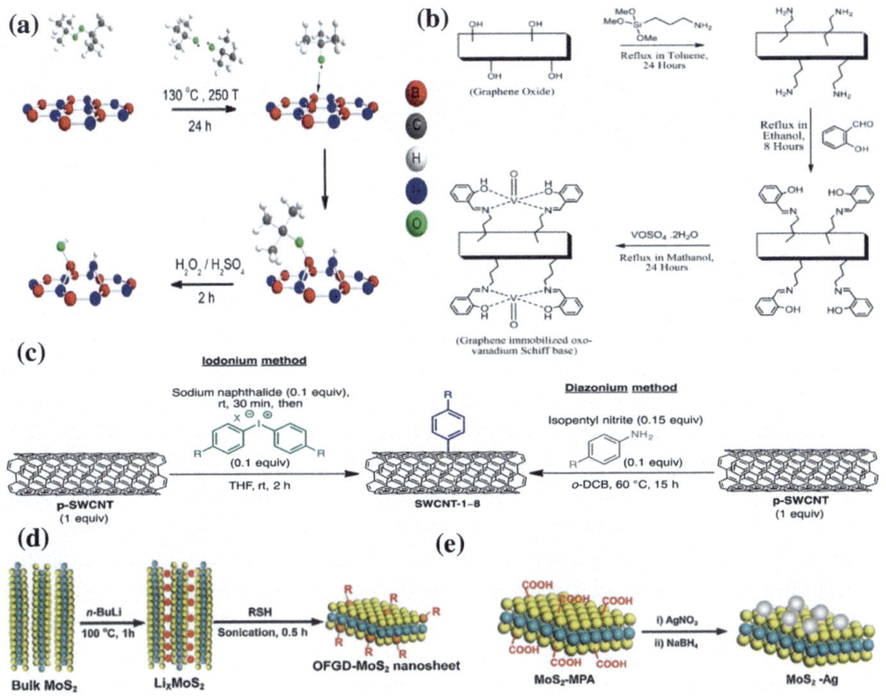

Fig. 2.15 Schematics show the covalent functionalisation of 2D nanomaterials for formation of **a** hydroxyl-functionalised boron nitride nanosheets. Reproduced with permission from [115]. Copyright 2012, the American Chemical Society; **b** graphene immobilised vanadium Schiff base. Reproduced with permision from [113]. Copyright 2011, the Royal Socity of Chemistry. **c** arene functionalised single-walled CNTs using iodonium and diazonium method. Reproduced with permission from [116]. Copyright 2016, the American Chemical Society; **d** OFGD-MoS_2 nanosheets (here R = MPA, TG, and CY). Reproduced with permission from [117]. Copyright 2014, the Royal Society of Chemistry; and **e** covalently linked Ag NPs on the surface of mercaptopropionic acid (MPA) functionalised MoS_2. Reproduced with permission from [117]. Copyright 2014, the Royal Society of Chemistry

unsymmetrical iodonium salts (Fig. 2.15c). A facile approach to synthesize organic functional group decorated (OFGD) MoS_2 nanosheets using thiol reagents as ligands was developed [117]. Thiol ligands such as mercaptopropionic acid (MPA), 1-thioglycerol (TG), and L-cysteine (CY) were used to modify MoS_2 by reaction with freshly prepared Li_xMoS_2 (Fig. 2.15d). A MoS_2-Ag nanocomposite was easily prepared by adding $AgNO_3$ to a solution of MPA-MoS_2 and sodium borohydride. Covalent functionalized MoS_2 nanosheets have a high density of nucleation sites due to the presence of –COOH groups (Fig. 2.15e).

Likewise, magnetic luminescent multifunctional NPs were synthesized using carboxyl-functionalized magnetic Fe_3O_4 NPs and amino-functionalized-silica-coated $NaYF_4$:Yb, Er fluorescent up-conversion NPs [118]. Magnetic luminescent multifunctional NPs were then used for biolabeling and fluorescent imaging of cancer cells.

The introduction of –COOH, –OH, and $–NH_2$ groups on nanomaterials can improve dispersibility and processability. Pristine nanomaterials have already shown exciting properties and diverse applications, while functionalized nanomaterials promise excellent possibilities as new candidates as they provide conventional platforms for tuning the structures and properties.

2.6.2 Noncovalent Functionalization

Attraction and repulsive forces between molecules are vital for supramolecular or noncovalent interactions and functionalization. Guest molecules can be adsorbed on the nanomaterials by chemisorption and/or physisorption. Noncovalent interactions can easily be found in many synthetic and natural systems used for recognition interaction, and detection [119]. Some forms of noncovalent interactions are π–π interactions, hydrogen bonding, polymer wrapping, electron donor-acceptor ligand systems, and weak interactions (van der Waals interactions) [120]. Although the energies of noncovalent interactions are less than those of covalent bonds, the overall effects over large interfaces are similar. The exciting feature of noncovalent functionalization is its reversibility or kinetic lability, resulting in an auto-correcting system. Using these interactions, systems can be designed with excellent response towards external stimuli (physical or chemical), such as pH and temperature. This has become the working principle for drug delivery using hydrogels [121]. Weak forces, such as van der Waals interactions, hydrogen bonding, and π–π interactions are effective in close proximity and easily accessible at room temperature. These forces easily influence surface interactions (hydrophilic and hydrophobic) and solvation of nanomaterials. For example, a high density of NPs in dispersion of 2D nanomaterials including CNTs, graphene, and h-BN can be achieved using solvents such as *N*-methylpyrrolidone (NMP), phenylethyl alcohol and *N,N*-dimethylformamide. During ultrasonication, NMP molecules can enter into nanosheets and adsorb to the surface. Subsequently, the NMP molecules can be easily removed by simply heating the sample at 340 °C as the NMP was physically adsorbed via van der Waals interactions on the surfaces [122]. Unlike covalent functionalization, noncovalent interactions do not disturb sp^2 carbon networks and result in nanomaterials with novel properties while maintaining some inherent properties. Overall, noncovalent functionalization can produce nanomaterials with improved reactivity, dispersibility, binding capacity, biocompatibility, catalytic activity, and sensing behavior [86, 123]. Hsiao et al. demonstrated the preparation of rGO/water-soluble polyurethane (WPU) nanocomposites using noncovalent functionalization for application in electromagnetic shielding [124]. In their work, GO was firstly functionalized using a stearyl trimethyl ammonium chloride surfactant to increase dispersibility and add ammonium moieties on the surface. Sulfonate-group-functionalized polyurethane (water soluble) was used as the matrix for the nanocomposite. A homogenous dispersion was observed in the nanocomposite due to the highly compatible electrostatic interaction between hydrophilic sulfonate

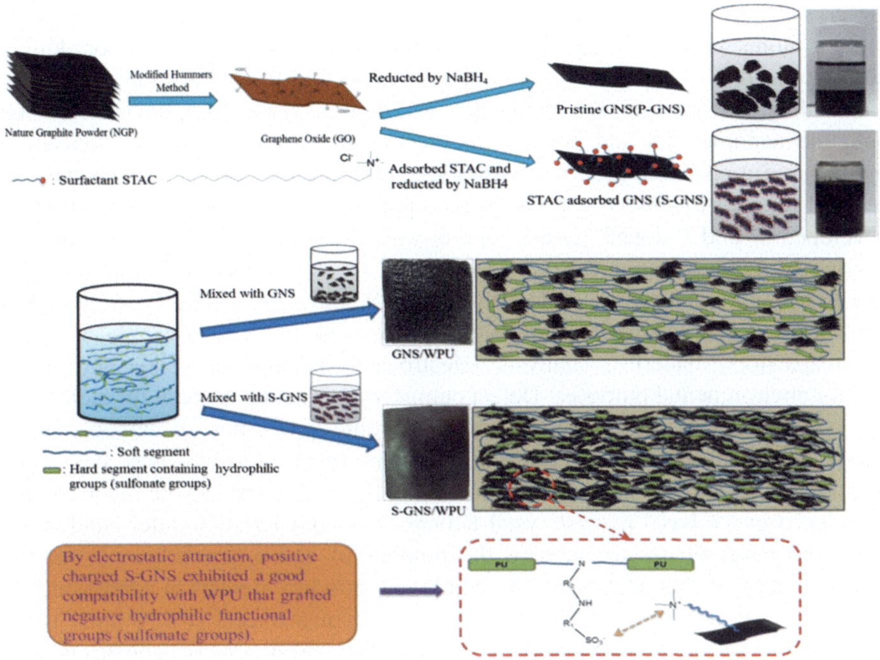

Fig. 2.16 Formation of nanocomposite using water-soluble polyurethane and cationic surfactants functionalised reduced GO. Reproduced with permission from [124]. Copyright 2013, Elsevier Science Ltd

groups of WPU and tertiary amine moieties present on the rGO (Fig. 2.16). In another report, a hydrogel nanocomposite of nickel sulfide NPs and gum karaya (GK) polysaccharide was prepared using non-covalent interactions for adsorption of organic dye from synthetic water [125]. Various bio-nanocomposites were successfully synthesized by non-covalent interaction between polylactic acid and different types of nanofillers [126].

2.6.3 Intrinsic Surface Engineering: Heteroatom Incorporation and Defect Engineering

Intrinsic surface engineering refers to modifications at the atomic level in the crystal structure of nanomaterials. It can be achieved by heteroatom incorporation or defect engineering. Incorporation or doping of electron-withdrawing and electron-donating elements (N, S, O, P, B, and other metals) changes the electronic, catalytic and structural properties of nanomaterials [125]. Dopants are also responsible for distortion of the crystal lattice and deviations in the band angles and lengths, resulting in the creation of many active sites. Dopants can be introduced in the crystal lattice by substitution of cations and anions or covalent bonding with foreign atoms [128]. Various synthesis routes have been developed for doping heteroatoms into nanomaterials via both in situ and post-treatment strategies. CVD, ball milling, and melt mixing approaches can be used for in situ preparation using both the nanomaterial and heteroatom precursors. However, thermal annealing/pyrolysis, wet chemical routes, arc-discharge- and plasma-assisted approaches can be used for post-treatment heteroatom incorporation [128]. For example, Yu et al. developed S- and P-doped 3D macroporous graphene aerogel using physical activation and reduction approaches for supercapacitor applications [129]. Phytic acid and thioglycolic acid were used as the P and S source, respectively. Single- and multi-dopant nanomaterials have been synthesized for diverse applications in supercapacitors, batteries, catalysis (electrocatalysis/photocatalysis), phosphors, and for environmental purposes. Defect engineering is also beneficial for modifying the structural, optical, biological, and electronic properties of nanomaterials. Although vacancies, defects, or trap states are often invisible, they contribute greatly to the surface activity of a material [84]. Defects can be intentionally introduced in the basal planes/crystal lattices of nanomaterials to alter band structures and basal planes to increase the number of active sites/edges. Primarily, physical and wet chemical routes are employed to induce defects in nanomaterials. Wang et al. reported defect-rich 2D WO_3 nanosheets ($\sim$10 nm) using template-assisted surfactant-based hydrothermal synthesis. WO_3 nanosheets with oxygen vacancies showed a high selectively for N_2 gas at 140 °C [130].

2.7 Nanoparticle–Polymer Composite Applications

Functionalized NPs have been dispersed in different polymer matrices to develop various types of novel polymer nanocomposites (PNCs) targeting superior efficacy and new functionalities for lightweight applications. PNCs have the benefits of the polymer matrix as well as the special characteristics of the nanoreinforcement phase. The addition of nanomaterials to polymers can achieve improved mechanical, electrical, thermal, optical, and biological properties. CNTs, nanoclays, GO, metal oxides NPs, metal NPs, nanofibers, QDs, and others have been used to prepare PNCs, where their lightweight applications include automotive, aerospace, packaging, construction, electrical, and electronics. Many available products in the market based on PNCs are benefitting from the growth and popularity of this field. PNCs have been successfully used as gas barriers for plastics bottles, sports items, and packaging. For example, clay (vermiculite)–polymer (butyl rubber) nanocomposites were used to develop commercial Wilson tennis balls (carrying double core). The PNC double core in this tennis ball allows it to maintain internal pressure for a long period of time and also acts as a gas barrier, increasing the shelf life [132]. The company Honeywell has introduced the first plastic bottles (especially for beer storage) using a nanoclay–polymer [polyethylene terephthalate (PET)] nanocomposite. This product offer many benefits, including a long shelf life (6 months), high

Table 2.2 Summarized some recent examples of PNCs that used for various applications

NPs	Polymer	Method used	Applications	Reference
CeO_2	PVP Poly(4vp-*co*-dvb) Polyacrylate	Electro-spinning Solution mixing Emulsion	Surface acoustic wave (SAW) devices Catalysis Solar cells	[136] [137] [138]
ZnO QDs $ZnO:Tb^{3+}$	PMMA and PCEM PAEK Polystyrene	ATRP Ball milling, hot pressing Wet chemical route	Thermal properties Electronic Solid-state lighting	[139] [140] [141]
TiO_2 Hydrogel	HDPE PVA PVA/PVP	Doctor Blade method Solution casting Sol-gel method	Reverse wettability, self-cleaning Antimicrobial activity Articular cartilage tissue regeneration	[142] [143] [144]
GO	PDMS/PGMA Chitosan Polyaniline	ATRP Wet chemical route Dilute polymerisation	Electronic and engineering Wastewater purification Energy storage	[145] [146] [147]
Clays	PP/TPU PLA + Ag NPs Chitosan/PVA	Blending Melt-compounding blending	Flame retardant Antibacterial Food packaging	[148] [149] [150]
CNTs	PVA + PEI PLA PEDOT:PSS	Solution casting Ball mill mixing Solution mixing	Antibacterial EMI shielding Thermoelectric	[151] [152] [153]
MoS_2	PAN	Intercalation	Li-ion batteries	[154]
NiS/ Ni_3S_4 hydrogel	Polysaccharide (GK) + PAAM	Free radical grafting + co-precipitation	Water purification	[125]
CdSe	PMMA	Blending	Memory device	[155]
Au	PEDOT:PSS	Solution mixing	Fuel cell	[156]
Ag	PE	Melt-compounding	Antibacterial	[157]

Abbreviations *PVP* (Polyvinylpyrrolidone), poly(4vp-co-dvb)[divinylbenzene (dvb) crosslinked 4-vinyl pyridine (4vp) polymer], *PMMA* [Poly(methyl methacrylate)], *PCEM* [Poly(2-(carbazol-9-yl)ethyl methacrylate)] *ATRP* (atom transfer radical polymerization), *PAEK* [poly (aryletherketon)], *HDPE* (high-density polyethylene), *PVA*(Polyvinyl alcohol), *PDMS* (polydimethylsiloxane), *PGMA* [poly(glycidyl methacrylate)], *PP* (polypropylene), *TPU* (thermoplastic polyurethane), *PLA* (polylactic acid), *PEI* (polyethyleneimine), *EMI* (electromagnetic interference), *PEDOT:PSS* [poly(3,4-ethylenedioxythiophene):poly (styrenesulfonate)], *PAAM* (polyacrylamide), *PAN* [(poly(acrylonitrile)], *PE* (polyethylene), *GK* (gum karaya)

beer colloidal stability, less oxygen permeation, and better taste (due to decreased light exposure) [132]. A PNC of rubber poly(dimethyl siloxane) and SiO_2 NPs was employed to prepare golf balls [133, 134]. A PNC based on clay/polyurethane showed excellent flame retardant behavior and has been commercialized for several applications, such as automobile seats, wire jackets (cables), textile cloths, packing films, paints, ablative materials for rocket cores, and coatings for stainless steel products [135]. Some recent examples from the literature of PNCs designed for biomedical, environmental, energy, and other applications are listed in Table 2.2.

In conclusion, various nanomaterials have found applications in nanocomposite materials. Due to the wide range of applications and emerging commercialized products, the market for PNCs is expected to exceed \$5.1 billion in 2020. Meanwhile, the market for metal oxide NPs and metal NPs could accrue a revenue of \$51 billion by 2026 due to their widespread application in sunscreens, cosmetics, diagnostics, bioimaging, sensors, and catalysis. QDs have been successfully commercialized for display, solid state lighting, and bioimaging applications. Many technology giants (e.g., Sony and Samsung) have already developed flat-screen TVs using QD technology to enhance color variations. The QD market is also expected to grow to \$6.0 billion by 2020 [158]. If these predictions are even close to reality, the world will witness several interesting technologies, i.e., nanoproducts/devices based on PNCs and other nanomaterials in the coming years.

2.8 Summary

The rapid progress in nanotechnology in recent years has shown several unprecedented discoveries that demonstrate its particular potential for a range of technological applications. Nanomaterials are at the core of the emerging nanotechnology field. Many highly motivated researchers are significantly contributing to the development of various nanomaterials and facile synthesis routes, exploring size-dependent properties, and investigating appropriate applications for solving our daily life challenges. This chapter focused on the classification of the different types of available nanomaterials, various common synthesis routes, several available surface modification/functionalization approaches, and the application of nanomaterials in polymer composites. Nanomaterials are classified based on their basic chemistry (inorganic, organic, or carbon nanomaterials). Furthermore, they are also categorized based on composition, morphology, dimensionality, agglomerated states, and uniformity. The advantages and disadvantages of various nanomaterial synthesis methods were discussed. Furthermore, functionalization strategies (covalent, noncovalent, heteroatom incorporation, and defect engineering) were introduced in detail. Recent research on the role of nanomaterial in polymers was reviewed. The discussed examples clearly demonstrated the importance of PNCs in end-user products and provided a glimpse into their widespread application in the near future. New nanomaterials, effects, concepts, and models to explain the observed experimental results are being developed regularly. Although nanomaterials have found varied applications, there are many concerns regarding environmental and health hazards due to their very small size and disposal in the natural environment. Thus, it is imperative to develop facile and environmentally friendly routes to synthesize nanomaterials. Overall, smart nanomanufacturing will be a crucial step in achieving commercial nanoproducts based on nanomaterials.

Acknowledgements The authors would like to thank the Department of Science and Technology and the Council for Scientific and Industrial Research, South Africa, for financial support.

References

1. Rai S, Rai A. Nanotechnology-the secret of fifth industrial revolution and the future of next generation. Jurnal Nasional 2017;7(2):61–6.
2. Kagan CR, Murray CB. Charge transport in strongly coupled quantum dot solids. Nat Nanotechnol. 2015;10(12):1013.
3. Scher JA, Elward JM, Chakraborty A. Shape matters: effect of 1D, 2D, and 3D isovolumetric quantum confinement in semiconductor nanoparticles. J Phys Chem C. 2016;120(43):24999–5009.
4. Pitkethly MJ. Nanomaterials—the driving force. Mater Today. 2004;7(12):20–9.
5. Sanchez C, et al. Applications of advanced hybrid organic–inorganic nanomaterials: from laboratory to market. Chem Soc Rev. 2011;40(2):696–753.
6. Lohse SE, Murphy CJ. Applications of colloidal inorganic nanoparticles: from medicine to energy. J Am Chem Soc. 2012;134(38):15607–20.
7. Nam KT, et al. Virus-enabled synthesis and assembly of nanowires for lithium ion battery electrodes. Science 2006;312(5775):885–8.
8. Hossein-Babaei F, et al. A resistive gas sensor based on undoped p-type anatase. Sens Actuators B Chem. 2005;110(1):28–35.
9. Ozgit D, Hiralal P, Amaratunga G. Flexible energy storage devices based on nanomaterials. In: 2016 IEEE 16th international conference on nanotechnology (IEEE-NANO). IEEE; 2016.
10. Fujishima A, Zhang X. Titanium dioxide photocatalysis: present situation and future approaches. C R Chim. 2006;9(5–6):750–60.
11. Activ P. Self-cleaning glass. 2008.
12. Serpone N, Dondi D, Albini A. Inorganic and organic UV filters: their role and efficacy in sunscreens and suncare products. Inorg Chim Acta. 2007;360(3):794–802.
13. Bera A, Belhaj H. Application of nanotechnology by means of nanoparticles and nanodispersions in oil recovery—a comprehensive review. J Nat Gas Sci Eng. 2016;34:1284–309.
14. áJames McQuillan A, Martin J. Characterisation and activity of sol–gel-prepared TiO_2 photocatalysts modified with Ca, Sr or Ba ion additives. J Mater Chem. 2000;10(10): 2358–63.
15. Kumar N, et al. Morphogenesis of ZnO nanostructures: role of acetate (COOH–) and nitrate (NO_3^-) ligand donors from zinc salt precursors in synthesis and morphology dependent photocatalytic properties. RSC Adv. 2015;5(48):38801–9.
16. Fernando RH, Nanocomposite and nanostructured coatings: Recent advancements. Washington: ACS Publications; 2009.
17. Domun N, et al. Improving the fracture toughness and the strength of epoxy using nanomaterials—a review of the current status. Nanoscale. 2015;7(23):10294–329.

18. Mout R, et al. Surface functionalization of nanoparticles for nanomedicine. Chem Soc Rev. 2012;41(7):2539–44.
19. Chakraborty I, Pradeep T. Atomically precise clusters of noble metals: emerging link between atoms and nanoparticles. Chem Rev. 2017;117(12):8208–71.
20. Zaleska-Medynska A, et al. Noble metal-based bimetallic nanoparticles: the effect of the structure on the optical, catalytic and photocatalytic properties. Adv Coll Interface Sci. 2016;229:80–107.
21. Yin H, Tang Z. Ultrathin two-dimensional layered metal hydroxides: an emerging platform for advanced catalysis, energy conversion and storage. Chem Soc Rev. 2016;45(18):4873–91.
22. Bromley R, Murray R, Yoffe A. The band structures of some transition metal dichalcogenides. III. Group VIA: trigonal prism materials. J Phys C Solid State Phys. 1972;5(7):759.
23. Wilson JA, Yoffe A. The transition metal dichalcogenides discussion and interpretation of the observed optical, electrical and structural properties. Adv Phys. 1969;18(73):193–335.
24. Xu N, Xu Y, Zhu J, Topological insulators for thermoelectrics. npj Quantum Mater. 2017; 2(1):51.
25. Butler SZ, et al. Progress, challenges, and opportunities in two-dimensional materials beyond graphene. ACS Nano. 2013;7(4):2898–926.
26. Duerloo K-AN, Li Y, Reed EJ. Structural phase transitions in two-dimensional Mo-and W-dichalcogenide monolayers. Nat. Commun. 2014;5:4214.
27. Venkata Subbaiah Y, Saji K, Tiwari A. Atomically thin MoS_2: a versatile nongraphene 2D material. Adv Funct Mater. 2016;26(13):2046–69
28. Lipatov A, et al. Few-layered titanium trisulfide (TiS_3) field-effect transistors. Nanoscale. 2015;7(29):12291–6.
29. Zolyomi V, Drummond N, Fal'ko V. Band structure and optical transitions in atomic layers of hexagonal gallium chalcogenides. Phys Rev B. 2013;87(19):195403.
30. Gonzalez JM, Oleynik II. Layer-dependent properties of SnS_2 and $SnSe_2$ two-dimensional materials. Phys Rev B. 2016;94(12):125443.
31. Verzhbitskiy I, Eda G. Chalcogenide nanosheets: optical signatures of many-body effects and electronic band structure. In: Inorganic nanosheets and nanosheet-based materials. Berlin: Springer; 2017. p. 133–62.
32. Shirasaki Y, et al. Emergence of colloidal quantum-dot light-emitting technologies. Nat Photonics. 2013;7(1):13.
33. Sun L, et al. Electrogenerated chemiluminescence from PbS quantum dots. Nano Lett. 2008;9(2):789–93.
34. Pietryga JM, et al. Spectroscopic and device aspects of nanocrystal quantum dots. Chem Rev. 2016;116(18):10513–622.
35. Zhao W, et al. Lattice dynamics in mono-and few-layer sheets of WS_2 and WSe_2. Nanoscale. 2013;5(20):9677–83.
36. Geim AK, Grigorieva IV. Van der Waals heterostructures. Nature. 2013;499(7459):419.
37. Prasad PN. Nanophotonics. Hoboken: Wiley; 2004.
38. Chang Y-M, et al. Enhanced free exciton and direct band-edge emissions at room temperature in ultrathin ZnO films grown on Si nanopillars by atomic layer deposition. ACS Appl Mater Interfaces. 2011;3(11):4415–9.
39. Zhao N, Qi L. Low-temperature synthesis of star-shaped PbS nanocrystals in aqueous solutions of mixed cationic/anionic surfactants. Adv Mater. 2006;18(3):359–62.
40. Kayanuma Y. Quantum-size effects of interacting electrons and holes in semiconductor microcrystals with spherical shape. Phys Rev B. 1988;38(14):9797.
41. Wise FW. Lead salt quantum dots: the limit of strong quantum confinement. Acc Chem Res. 2000;33(11):773–80.
42. Bellessa J, et al. Quantum-size effects on radiative lifetimes and relaxation of excitons in semiconductor nanostructures. Phys Rev B. 1998;58(15):9933.
43. Xu G, et al. New generation cadmium-free quantum dots for biophotonics and nanomedicine. Chem Rev. 2016;116(19):12234–327.

44. Chen X, Liu Y, Ma Q. Recent advances in quantum dot-based electrochemiluminescence sensors. J Mater Chem C 2018;6:942.
45. Kairdolf BA, et al. Semiconductor quantum dots for bioimaging and biodiagnostic applications. Ann Rev Anal Chem. 2013;6:143–62.
46. Xu Y, et al. Recent progress in two-dimensional inorganic quantum dots. Chem Soc Rev. 2018;47:586.
47. Ealia SAM, Saravanakumar, M. A review on the classification, characterisation, synthesis of nanoparticles and their application. In: IOP conference series: materials science and engineering. IOP Publishing; 2017.
48. Akbarzadeh A, et al. Liposome: classification, preparation, and applications. Nanoscale Res Lett. 2013;8(1):102.
49. Orive G, et al. Biomaterials for promoting brain protection, repair and regeneration. Nat Rev Neurosci. 2009;10(9):682.
50. Adair JH, et al. Nanoparticulate alternatives for drug delivery. ACS Nano. 2010;4(9): 4967–70.
51. Rao JP, Geckeler KE. Polymer nanoparticles: preparation techniques and size-control parameters. Prog Polym Sci. 2011;36(7):887–913.
52. Georgakilas V, et al. Broad family of carbon nanoallotropes: classification, chemistry, and applications of fullerenes, carbon dots, nanotubes, graphene, nanodiamonds, and combined superstructures. Chem Rev. 2015;115(11):4744–822.
53. Gogotsi Y, Presser V. Carbon nanomaterials. Boca Raton: CRC Press; 2013.
54. Gogotsi Y. Not just graphene: the wonderful world of carbon and related nanomaterials. MRS Bull. 2015;40(12):1110–21.
55. Kroto H, et al. This week's citation classic®. Nature. 1985;318:162–3.
56. Iijima S. Helical microtubules of graphitic carbon. Nature. 1991;354(6348):56.
57. Boehm H, et al. Dünnste kohlenstoff-folien. Z Naturforsch B. 1962;17(3):150–3.
58. Novoselov KS, et al. Electric field effect in atomically thin carbon films. Science 2004; 306(5696):666–9.
59. Verma S, et al. Graphene oxide: an efficient and reusable carbocatalyst for aza-Michael addition of amines to activated alkenes. Chem Commun. 2011;47(47):12673–5.
60. Ong W-J, et al. Graphitic carbon nitride (g-C_3N_4)-based photocatalysts for artificial photosynthesis and environmental remediation: are we a step closer to achieving sustainability? Chem Rev. 2016;116(12):7159–329.
61. Buzea C, Pacheco II, Robbie K. Nanomaterials and nanoparticles: sources and toxicity. Biointerphases 2007;2(4):MR17–71.
62. Pokropivny V, Skorokhod V. Classification of nanostructures by dimensionality and concept of surface forms engineering in nanomaterial science. Mater Sci Eng C. 2007;27(5–8):990–3.
63. Tiwari JN, Tiwari RN, Kim KS. Zero-dimensional, one-dimensional, two-dimensional and three-dimensional nanostructured materials for advanced electrochemical energy devices. Prog Mater Sci. 2012;57(4):724–803.
64. Tan C, et al. Recent advances in ultrathin two-dimensional nanomaterials. Chem Rev. 2017;117(9):6225–331.
65. Jiang P, et al. Emission-tunable near-infrared Ag_2S quantum dots. Chem Mater. 2011;24(1): 3–5.
66. Duan C, et al. Controllability of the Coulomb charging energy in close-packed nanoparticle arrays. Nanoscale. 2013;5(21):10258–66.
67. Zhang J, et al. Nonepitaxial growth of hybrid core-shell nanostructures with large lattice mismatches. Science. 2010;327(5973):1634–8.
68. Jeon S, et al. Fullerene embedded shape memory nanolens array. Sci Rep. 2013;3:3269.
69. Shokuhfar T, et al. Intercalation of anti-inflammatory drug molecules within TiO_2 nanotubes. Rsc Adv. 2013;3(38):17380–6.
70. Zhu J, Xue D. Crystallography and interfacial kinetic controlled ultra-uniform single crystal silver nanobelts and their optical properties. CrystEngComm. 2014;16(4):642–8.

71. Kumar N, et al. Controlled synthesis of microsheets of ZnAl layered double hydroxides hexagonal nanoplates for efficient removal of Cr(VI) ions and anionic dye from water. J Environ Chem Eng. 2017;5(2):1718–31.
72. Nayak BB, Behera D, Mishra BK. Synthesis of silicon carbide dendrite by the arc plasma process and observation of nanorod bundles in the dendrite arm. J Am Ceram Soc. 2010; 93(10):3080–3.
73. Bruno E, et al. Comparison of the Sensing Properties of ZnO nanowalls-based sensors toward low concentrations of CO and NO_2. Chemosensors. 2017;5(3):20.
74. Karakouz T, et al. Morphology and refractive index sensitivity of gold island films. Chem Mater. 2009;21(24):5875–85.
75. Hirahara K, Nakayama Y. The effect of a tin oxide buffer layer for the high yield synthesis of carbon nanocoils. Carbon. 2013;56:264–70.
76. Choudhury BD, et al. Silicon nanopillar arrays with SiO_2 overlayer for biosensing application. Opt Mater Express. 2014;4(7):1345–54.
77. Pan A, et al. Template-free synthesis of VO_2 hollow microspheres with various interiors and their conversion into V_2O_5 for Lithium-Ion batteries. Angew Chem. 2013;125(8):2282–6.
78. Biswas A, et al. Advances in top–down and bottom–up surface nanofabrication: techniques, applications & future prospects. Adv Coll Interface Sci. 2012;170(1–2):2–27.
79. Zhou Y, et al. Top-down preparation of active cobalt oxide catalyst. ACS Catal. 2016;6 (10):6699–703.
80. Priyadarshana G, et al. Synthesis of magnetite nanoparticles by top-down approach from a high purity ore. J Nanomater. 2015;16(1):317.
81. Wang Y, Xia Y. Bottom-up and top-down approaches to the synthesis of monodispersed spherical colloids of low melting-point metals. Nano Lett. 2004;4(10):2047–50.
82. Gao M-R, et al. Nanostructured metal chalcogenides: synthesis, modification, and applications in energy conversion and storage devices. Chem Soc Rev. 2013;42(7):2986–3017.
83. Liang HW, et al. Macroscopic-scale template synthesis of robust carbonaceous nanofiber hydrogels and aerogels and their applications. Angew Chem Int Ed. 2012;51(21):5101–5.
84. Kumar N, et al. Sustainable one-step synthesis of hierarchical microspheres of PEGylated MoS_2 nanosheets and MoO_3 nanorods: their cytotoxicity towards lung and breast cancer cells. Appl Surf Sci. 2017;396:8–18.
85. Jiang T, et al. Hydrothermal synthesis of Ag@MSiO_2@Ag three core–shell nanoparticles and their sensitive and stable SERS properties. Nanoscale. 2016;8(9):4908–14.
86. Kumar N, Ray SS, Ngila JC. Ionic liquid-assisted synthesis of Ag/Ag_2Te nanocrystals via a hydrothermal route for enhanced photocatalytic performance. New J Chem. 2017;41(23): 14618–26.
87. Kumar N, et al. Controlled microstructural hydrothermal synthesis of strontium selenides host matrices for Eu^{II} and Eu^{III} luminescence. Mater Lett. 2015;146:51–4.
88. Kumar N, et al. A novel approach to low-temperature synthesis of cubic HfO_2 nanostructures and their cytotoxicity. Sci Rep. 2017;7(1):9351.
89. Lu Q, Gao F, Zhao D. One-step synthesis and assembly of copper sulfide nanoparticles to nanowires, nanotubes, and nanovesicles by a simple organic amine-assisted hydrothermal process. Nano Lett. 2002;2(7):725–8.
90. Solanki JN, Murthy ZVP. Controlled size silver nanoparticles synthesis with water-in-oil microemulsion method: a topical review. Ind Eng Chem Res. 2011;50(22):12311–23.
91. Dhand C, et al. Methods and strategies for the synthesis of diverse nanoparticles and their applications: a comprehensive overview. Rsc Adv. 2015;5(127):105003–37.
92. Mandal M, et al. Micelle-mediated UV-photoactivation route for the evolution of Pdcore-Aushell and Pdcore-Agshell bimetallics from photogenerated Pd nanoparticles. J Photochem Photobiol A. 2004;167(1):17–22.
93. Martínez-Rodríguez RA, et al. Synthesis of Pt nanoparticles in water-in-oil microemulsion: effect of HCl on their surface structure. J Am Chem Soc. 2014;136(4):1280–3.

94. Yıldırım ÖA, Durucan C. Synthesis of zinc oxide nanoparticles elaborated by microemulsion method. J Alloy Compd. 2010;506(2):944–9.
95. Akhtar MS, Panwar J, Yun Y-S. Biogenic synthesis of metallic nanoparticles by plant extracts. ACS Sustain Chem Eng. 2013;1(6):591–602.
96. Shankar SS, et al. Rapid synthesis of Au, Ag, and bimetallic Au core–Ag shell nanoparticles using Neem (*Azadirachta indica*) leaf broth. J Colloid Interface Sci. 2004;275(2):496–502.
97. Maensiri S, et al. Indium oxide (In_2O_3) nanoparticles using Aloe vera plant extract: synthesis and optical properties. J Optoelectron Adv Mater. 2008;10:161–5.
98. Coccia F, et al. One-pot synthesis of lignin-stabilised platinum and palladium nanoparticles and their catalytic behaviour in oxidation and reduction reactions. Green Chem. 2012;14 (4):1073–8.
99. Sathishkumar M, et al. Phyto-crystallization of palladium through reduction process using *Cinnamom zeylanicum* bark extract. J Hazard Mater. 2009;171(1–3):400–4.
100. Sintubin L, et al. Lactic acid bacteria as reducing and capping agent for the fast and efficient production of silver nanoparticles. Appl Microbiol Biotechnol. 2009;84(4):741–9.
101. Bharde A, et al. Extracellular biosynthesis of magnetite using fungi. Small. 2006;2(1): 135–41.
102. Bansal V, et al. Fungus-mediated biosynthesis of silica and titania particles. J Mater Chem. 2005;15(26):2583–9.
103. Kundu S, Nithiyanantham U. DNA-mediated fast synthesis of shape-selective ZnO nanostructures and their potential applications in catalysis and dye-sensitized solar cells. Ind Eng Chem Res. 2014;53(35):13667–79.
104. Charitidis CA, et al. Manufacturing nanomaterials: from research to industry. Manuf Rev. 2014;1:11.
105. Moon J-W, et al. Manufacturing demonstration of microbially mediated zinc sulfide nanoparticles in pilot-plant scale reactors. Appl Microbiol Biotechnol. 2016;100(18): 7921–31.
106. Riche CT, et al. Flow invariant droplet formation for stable parallel microreactors. Nat Commun. 2016;7:10780.
107. Liu C, Cheng H-M. Carbon nanotubes: controlled growth and application. Materials Today. 2013;16(1–2): 19–28.
108. Pirard SL, Douven S, Pirard J-P. Large-scale industrial manufacturing of carbon nanotubes in a continuous inclined mobile-bed rotating reactor via the catalytic chemical vapor deposition process. Front Chem Sci Eng. 2017;11(2): 280–9.
109. Sperling RA, Parak WJ. Surface modification, functionalization and bioconjugation of colloidal inorganic nanoparticles. Philos Trans R Soc Lond A Math Phys Eng Sci. 1915;2010(368):1333–83.
110. Treccani L, et al. Functionalized ceramics for biomedical, biotechnological and environmental applications. Acta Biomater. 2013;9(7):7115–50.
111. Stamov DR, et al. The impact of heparin intercalation at specific binding sites in telopeptide-free collagen type I fibrils. Biomaterials. 2011;32(30):7444–53.
112. Gusain R, et al. Reduced graphene oxide–CuO nanocomposites for photocatalytic conversion of CO_2 into methanol under visible light irradiation. Appl Catal B. 2016;181:352–62.
113. Mungse HP, et al. Grafting of oxo-vanadium Schiff base on graphene nanosheets and its catalytic activity for the oxidation of alcohols. J Mater Chem. 2012;22(12):5427–33.
114. Gusain R, et al. Covalently attached graphene–ionic liquid hybrid nanomaterials: synthesis, characterization and tribological application. J Mater Chem A. 2016;4(3):926–37.
115. Sainsbury T, et al. Oxygen radical functionalization of boron nitride nanosheets. J Am Chem Soc. 2012;134(45):18758–71.
116. He M, Swager TM. Covalent functionalization of carbon nanomaterials with iodonium salts. Chem Mater. 2016;28(23):8542–9.
117. Zhou L, et al. Facile approach to surface functionalized MoS_2 nanosheets. RSC Adv. 2014; 4(61):32570–8.

118. Mi C, et al. Multifunctional nanocomposites of superparamagnetic (Fe_3O_4) and NIR-responsive rare earth-doped up-conversion fluorescent ($NaYF_4$: Yb, Er) nanoparticles and their applications in biolabeling and fluorescent imaging of cancer cells. Nanoscale. 2010;2(7):1141–8.
119. Lehn J-M. Toward complex matter: supramolecular chemistry and self-organization. Proc Natl Acad Sci. 2002;99(8):4763–8.
120. Georgakilas V, et al. Noncovalent functionalization of graphene and graphene oxide for energy materials, biosensing, catalytic, and biomedical applications. Chem Rev. 2016; 116(9):5464–519.
121. Gupta P, Vermani K, Garg S. Hydrogels: from controlled release to pH-responsive drug delivery. Drug Discov Today. 2002;7(10):569–79.
122. Di Crescenzo A, Ettorre V, Fontana A. Non-covalent and reversible functionalization of carbon nanotubes. Beilstein J Nanotechnol. 2014;5:1675.
123. Gusain R, et al. Ionic-liquid-functionalized copper oxide nanorods for photocatalytic splitting of water. ChemPlusChem. 2016;81(5):489–95.
124. Hsiao S-T, et al. Using a non-covalent modification to prepare a high electromagnetic interference shielding performance graphene nanosheet/water-borne polyurethane composite. Carbon. 2013;60:57–66.
125. Kumar N, et al. Efficient removal of rhodamine 6G dye from aqueous solution using nickel sulphide incorporated polyacrylamide grafted gum karaya bionanocomposite hydrogel. RSC Adv. 2016;6(26):21929–39.
126. Sinha Ray S. Polylactide-based bionanocomposites: a promising class of hybrid materials. Acc Chem Res. 2012;45(10):1710–20.
127. Wang X, et al. Heteroatom-doped graphene materials: syntheses, properties and applications. Chem Soc Rev. 2014;43(20):7067–98.
128. Kong X-K, Chen C-L, Chen Q-W. Doped graphene for metal-free catalysis. Chem Soc Rev. 2014;43(8):2841–57.
129. Yu X, Kang Y, Park HS. Sulfur and phosphorus co-doping of hierarchically porous graphene aerogels for enhancing supercapacitor performance. Carbon. 2016;101:49–56.
130. Wang Z, Wang D, Sun J. Controlled synthesis of defect-rich ultrathin two-dimensional WO_3 nanosheets for NO_2 gas detection. Sens Actuators B Chem. 2017;245:828–34.
131. Sanchez C, et al. Applications of hybrid organic–inorganic nanocomposites. J Mater Chem. 2005;15(35–36):3559–92.
132. Ke Y, Stroeve P. Polymer-layered silicate and silica nanocomposites. Amsterdam: Elsevier; 2005.
133. Brinker CJ, Clark DE, Ulrich DR. Better ceramics through chemistry III. Pittsburgh, PA (USA): Materials Research Society; 1988.
134. Wen J, Mark JE. Sol–gel preparation of composites of poly (dimethylsiloxane) with SiO_2 and SiO_2/TiO_2, and their mechanical properties. Polym J. 1995;27(5):492.
135. Mago G, et al. Polymer nanocomposite processing, characterization, and applications. J Nanomat 2010;2010.
136. Liu Y, et al. Electrospun CeO_2 nanoparticles/PVP nanofibers based high-frequency surface acoustic wave humidity sensor. Sens Actuators B Chem. 2016;223:730–7.
137. Sabitha G, et al. Ceria/vinylpyridine polymer nanocomposite: an ecofriendly catalyst for the synthesis of 3, 4-dihydropyrimidin-2 (1H)-ones. Tetrahedron Lett. 2005;46(47):8221–4.
138. Aguirre M, et al. Hybrid acrylic/CeO_2 nanocomposites using hydrophilic, spherical and high aspect ratio CeO_2 nanoparticles. J Mater Chem A. 2014;2(47):20280–7.
139. Sato M, et al. Preparation and properties of polymer/zinc oxide nanocomposites using functionalized zinc oxide quantum dots. Eur Polym J. 2008;44(11):3430–8.
140. Vaishnav D, Goyal R. Thermal and dielectric properties of high performance polymer/ZnO nanocomposites. In: IOP conference series: materials science and engineering. IOP Publishing; 2014.

141. Prakash J, et al. Phosphor polymer nanocomposite: ZnO:Tb^{3+} embedded polystyrene nanocomposite thin films for solid-state lighting applications. ACS Appl Nano Mater. 2018;1:977.
142. Xu QF, et al. Superhydrophobic TiO_2–polymer nanocomposite surface with UV-induced reversible wettability and self-cleaning properties. ACS Appl Mater Interfaces. 2013;5(18): 8915–24.
143. Dhanasekar M, et al. Ambient light antimicrobial activity of reduced graphene oxide supported metal doped TiO_2 nanoparticles and their PVA based polymer nanocomposite films. Mater Res Bull. 2018;97:238–43.
144. Cao L, et al. Biocompatible nanocomposite of TiO_2 incorporated bi-polymer for articular cartilage tissue regeneration: a facile material. J Photochem Photobiol B. 2018;178:440–6.
145. Song S, Zhai Y, Zhang Y. Bioinspired graphene oxide/polymer nanocomposite paper with high strength, toughness, and dielectric constant. ACS Appl Mater Interfaces. 2016;8(45): 31264–72.
146. Fan J, Grande CD, Rodrigues DF. Biodegradation of graphene oxide-polymer nanocomposite films in wastewater. Environ Sci Nano. 2017;4(9):1808–16.
147. Xu J, et al. Hierarchical nanocomposites of polyaniline nanowire arrays on graphene oxide sheets with synergistic effect for energy storage. ACS Nano. 2010;4(9):5019–26.
148. Kannan M, Thomas S, Joseph K. Flame-retardant properties of nanoclay-filled thermoplastic polyurethane/polypropylene nanocomposites. J Vinyl Addit Technol. 2015;23:E72.
149. Cai S, Pourdeyhimi B, Loboa EG. High-throughput fabrication method for producing a silver-nanoparticles-doped nanoclay polymer composite with novel synergistic antibacterial effects at the material interface. ACS Appl Mater Interfaces. 2017;9(25):21105–15.
150. Zhang L, et al. Sodium lactate loaded chitosan-polyvinyl alcohol/montmorillonite composite film towards active food packaging. Innovative Food Sci Emerg Technol. 2017;42:101–8.
151. Wu W-T, et al. Fabrication of silver/cross-linked poly (vinyl alcohol) cable-like nanostructures under γ-ray irradiation. Nanotechnology. 2005;16(12):3017.
152. Chizari K, et al. Three-dimensional printing of highly conductive polymer nanocomposites for EMI shielding applications. Mater Today Commun. 2017;11:112–8.
153. Hsu J-H, et al. Origin of unusual thermoelectric transport behaviors in carbon nanotube filled polymer composites after solvent/acid treatments. Org Electron. 2017;45:182–9.
154. Santa Ana MA, et al. Poly (acrylonitrile)–molybdenum disulfide polymer electrolyte nanocomposite. J Mater Chem. 2006;16(30):3107–13.
155. Kim J-M, et al. Non-volatile organic memory based on CdSe nano-particle/PMMA blend as a tunneling layer. Synth Met. 2011;161(13–14):1155–8.
156. Zhang R-C, et al. Gold nanoparticle-polymer nanocomposites synthesized by room temperature atmospheric pressure plasma and their potential for fuel cell electrocatalytic application. Sci Rep. 2017;7:46682.
157. Mural PKS, et al. Porous membranes designed from bi-phasic polymeric blends containing silver decorated reduced graphene oxide synthesized via a facile one-pot approach. RSC Adv. 2015;5(41):32441–51.
158. Talapin DV, Shevchenko EV. Introduction: nanoparticle chemistry. Chem Rev. 2016; 116(18):10343–5.
159. Justyna C, Jacek H, Artur M, Marta M, Tomasz A, Kowalewski, et al. Synthesis of carbon nanotubes by the laser ablation method: effect of laser wavelength. Phys. Status Solidi(b). 2015;252(8):1860–7.

Chapter 3
A Brief Overview of Layered Silicates and Polymer/Layered Silicate Nanocomposite Formation

Suprakas Sinha Ray and Vincent Ojijo

Abstract Layered silicate-containing polymer nanocomposites attract great interest in today's advanced composite materials research because it is possible to achieve impressive property improvements when compared with neat polymers or conventional filler-filled composites. In its pristine form layered silicate is hydrophilic and not compatible with hydrophobic polymer matrices. To make layered silicate compatible with hydrophobic polymer matrix, one must convert hydrophilic surface to an organophilic one. This chapter briefly summarizes the structure and properties of pristine and organically modified layered silicates. This chapter also provides overview of layered silicate–containing polymer nanocomposites formation.

3.1 Introduction

3.1.1 Structure and Properties of Pristine Layered Silicates

Among the all minerals, silicates are the largest, the most interesting and the most complicated class of minerals by far. Approximately 30% of all minerals are silicates and some geologists estimate that 90% of the Earth's crust is made up of silicates [1]. With oxygen and silicon the two most abundant elements in the earth's crust silicates abundance is no real surprise. The basic chemical unit of silicates is the (SiO_4) tetrahedron shaped anionic group with a negative four charge (−4). The

S. Sinha Ray (✉) · V. Ojijo
DST-CSIR National Centre for Nanostructured Materials,
Council for Scientific and Industrial Research, Pretoria 0001, South Africa
e-mail: rsuprakas@csir.co.za

V. Ojijo
e-mail: vojijo@csir.co.za

S. Sinha Ray
Department of Applied Chemistry, University of Johannesburg,
Doornfontein 2028, Johannesburg, South Africa
e-mail: ssinharay@uj.ac.za

S. Sinha Ray (ed.), *Processing of Polymer-based Nanocomposites*,
Springer Series in Materials Science 277,
https://doi.org/10.1007/978-3-319-97779-9_3

central silicon ion has a charge of positive four while each oxygen has a charge of negative two (−2) and thus each silicon-oxygen bond is equal to one half (1/2) the total bond energy of oxygen. This condition leaves the oxygen's with the option of bonding to another silicon ion and therefore linking one (SiO_4) tetrahedron to another and another, etc. The complicated structures that these silicate tetrahedrons form are truly amazing. They can form as single units (known as nesosilicates), double units (known as sorosilicates), chains (known as inosilicates), sheets (known as phyllosilicates), rings (known as cyclosilicates) and framework structures (known as tectosilicates). The different ways that the silicate tetrahedrons combine is what makes the silicate class the largest, the most interesting and the most complicated class of minerals.

The most generally used silicates for the preparation of polymer nanocomposites (PNCs) belong to the general family of phyllosilicates, i.e., layered or sheets like structure or more commonly called as layered silicates (LSs) [2]. In this subclass, rings of tetrahedrons are linked by shared oxygens to other rings in a two dimensional plane that produces a sheet-like structure. The silicon to oxygen ratio is generally 1:2.5 (or 2:5) because only one oxygen is exclusively bonded to the silicon and the other three are half shared (1.5) to other silicons. The symmetry of the members of this group is controlled chiefly by the symmetry of the rings but is usually altered to a lower symmetry by other ions and other layers. The typical crystal habit of this subclass is therefore flat, platy, and book-like and always displays good basal cleavage. Table 3.1 listed some common members of the phyllosilicates.

The thickness of each layer or sheet is around 1 nm, and the lateral dimensions of these layers may vary from thirty nano meters to several microns or larger, depending on the particular LS. Stacking of the layers leads to a regular van der Waals gap between the layers called the *interlayer* or *gallery*. The layer surface has 0.25–0.9 negative charges per unit cell and various types of exchangeable cations within the interlayer galleries. Isomorphic substitution within the layers (for example, Al^{+3} replaced by Mg^{+2} or Fe^{+2}, or Mg^{+2} replaced by Li^{+1}) generates negative charges that are counterbalanced by alkali and alkaline earth cations situated inside the galleries. This type of LS is characterized by a moderate surface charge known as the cation exchange capacity (*CEC*), and generally expressed as mequiv/100 g. This charge is not locally constant, but varies from layer to layer, and must be considered as an average value over the whole crystal. LSs have two types of structure: tetrahedral-substituted and octahedral substituted. In the case of tetrahedrally substituted layered silicates the negative charge is located on the

Table 3.1 Some common members of phyllosilicates

Clay group	Chlorite, Glaoconite, Illite, Kaolinite, Montmorillonite, Palygorskite, Pyrophyllite, Saoconite, Talc, Vermiculate
Mica group	Biotite, Lepidolite, Muscovite, Paragonite, Phlogopite, Zinnwaldite
Sepentine group	Antigorite, Clinochrysotile, Lizardite, Orthochrysotile, serpentine
Others	Allophane, Apophyllite, Bannisterite, Cavansite, Chrysocolla, Delhayelite, Elpidite, Fedorite, etc.

surface of silicate layers, and hence, the polymer matrices can react/interact more readily with those than with octahedrally-substituted material.

Clays are the most important minerals within the LSs or phyllosilicates and generally contain huge percentages of trapped water molecules inside the silicate galleries. Most of them are chemically and structurally analogous to each other but contain varing amounts of water and allow more substitution of their cation. Generally, clays minerals are divided into three major groups [1]:

The kaolinite group. Kaolinite, dickite, and nacrite are the members of this group. The general chemical formula is $Al_2Si_2O_5(OH)_4$. All members have the same chemistry but different in structures. This property is generally known as polymorphs. The general structure of this group is composed of silicate sheets bonded to aluminum oxide/hydroxide layers called gibbsite layers [1].

The semctite group. This group is consisting of several LS minerals and most important are vermiculate, saponite, hectorite, montmorillonite (MMT), talc, sauconite, and nontronite. The general formula is (Ca, Na, H)(Al, Mg, Fe, Zn)$_2$(Si, Al)$_4O_{10}(OH)_{2-x}H_2O$, where x represents the variable amount of water that members of this group could contain. In this group the gibbsite layers of the kaolinite group is replaced by a similar layer known as oxide brucite, ($Mg_2(OH)_4$). The structure of this group is composed of silicate layers sandwiching a brucite layer in between and water molecules present in between the sandwich layers [1].

The illite or the mica-clay group. This group is basically a hydrated microscopic muscovite. Muscovite is a common rock forming mineral and is found in igneous, metamorphic and detrital sedimentary rocks. The general formula is (K, H)Al_2(Si, Al)$_4O_{10}(OH)_{2-x}H_2O$, where x represents the variable amount of water that members of this group could contain and the structure of this is similar to that of smectite group [1].

Among the three major groups, smectite types or more precisely MMT, saponite and hectorite are the most commonly used LSs in the field of polymer nanocomposite (PNC) technology. Their chemical formula and characteristic parameters are summarized in Table 3.2. Again, among montmorillonite, saponite, hectorite, montmorillonite (MMT) is the most commonly used layered silicates for the fabrication of PNCs, because it is highly abundant and inexpensive. MMT is the name given to the LS found near montmorillonite in France, where MMT was first identified by Knight in 1896.

Table 3.2 Chemical formula and characteristic parameter of most commonly used layered silicates

Layered silicates	Chemical formula	CEC (mequiv/100 g)	Particle length (nm)
Montmorillonite	$Mx(Al_{4-x}Mg_x)Si_8O_{20}(OH)_4$	110	100–150
Saponite	$M_xMg_6(Si_{8-x}Al_x)Si_8O_{20}(OH)_4$	87	50–60
Hectorite	$M_x(Mg_{6-x}Li_x)Si_8O_{20}(OH)_4$	120	200–300

M monovalent cation; *x* degree of isomorphous substitution (between 0.5 and 1.3)

Table 3.3 Physical characteristics of montmorillonite clay

Color	Usually white, gray or pink with tints of yellow or green
Luster	Dull
Transparency	Crystals are translucent and masses are opaque
Crystal system	Monoclinic; 2/m
Unit cell molecular weight	540.5 (g/ml)
Crystal habits	Usually found in compact or lamellar masses. Also seen as inclusions in quartz as fibers and powder-like masses
Field indicator	Softness and soapy feel
Cleavage	Perfect in one direction
Hardness	1–2 (in Moh's scale and at room temperature)
Average specific gravity	2.3–3 g/ml
Facture	Uneven to lamellar
Swelling behavior	MMT crystals swell almost 30 times their original volume when added to water
Notable occurrences	China, France, Italy, Japan, USA and many other localities worldwide
Associated minerals	Other clays, granite, biotite and quartz

Source Amethyst Galleries, Inc. web sites. © 2006 Amethyst Galleries, Inc.

The specific surface area of MMT is equal to 750−800 m^2/g and the modulus of each MMT sheet is around 250 GPa [3]. The interlayer thickness of hydrated MMT is equal to 1.45 nm and the average density ρ = 2.385 g/ml. Drying MMT at 150 °C reduces the gallery height to 0.28 nm which corresponds to a water monolayer and hence the interlayer spacing decreases to 0.94 nm and the average density increases to 3.138 g/ml. Various properties of MMT are tabulated in Table 3.3 and unit structure is presented in Fig. 3.1.

Although MMT is highly abundant and inexpensive, however, it is a mineral with variable composition, which make impossible to purify MMT completely. For this reason, there is a growing interest to use fully or semi-synthetic LSs for the preparation of PNCs, because they have well-controlled physical and chemical properties. One of the most commonly used synthetic LSs in PNC technology is synthetic fluorine mica (SFM). It is generally synthesized by heating a mixer of talc and Na_2SiF_6 for several hours in an electric furnace [4]. Like MMT, SFM also belongs to the same generally family of 2:1 layered or phyllosilicates. The only difference between MMT and SFM is that SFM $[NaMg_{2.5}(Si_4O_{10})F_2]$ contains 'F' groups on its surface. Figure 3.2 represents the unit structure of SFM and physical characteristics of natural mica are summarized in Table 3.4.

Two particular characteristics of LSs are generally considered for the preparation of PNCs. The first is the ability of the silicate particles to disperse into individual layers. The second characteristic is the ability to fine-tune their surface chemistry through ion exchange reactions with organic and inorganic cations. These two characteristics are, of course, interrelated since the degree of dispersion of LS in a particular polymer matrix depends on the interlayer cation.

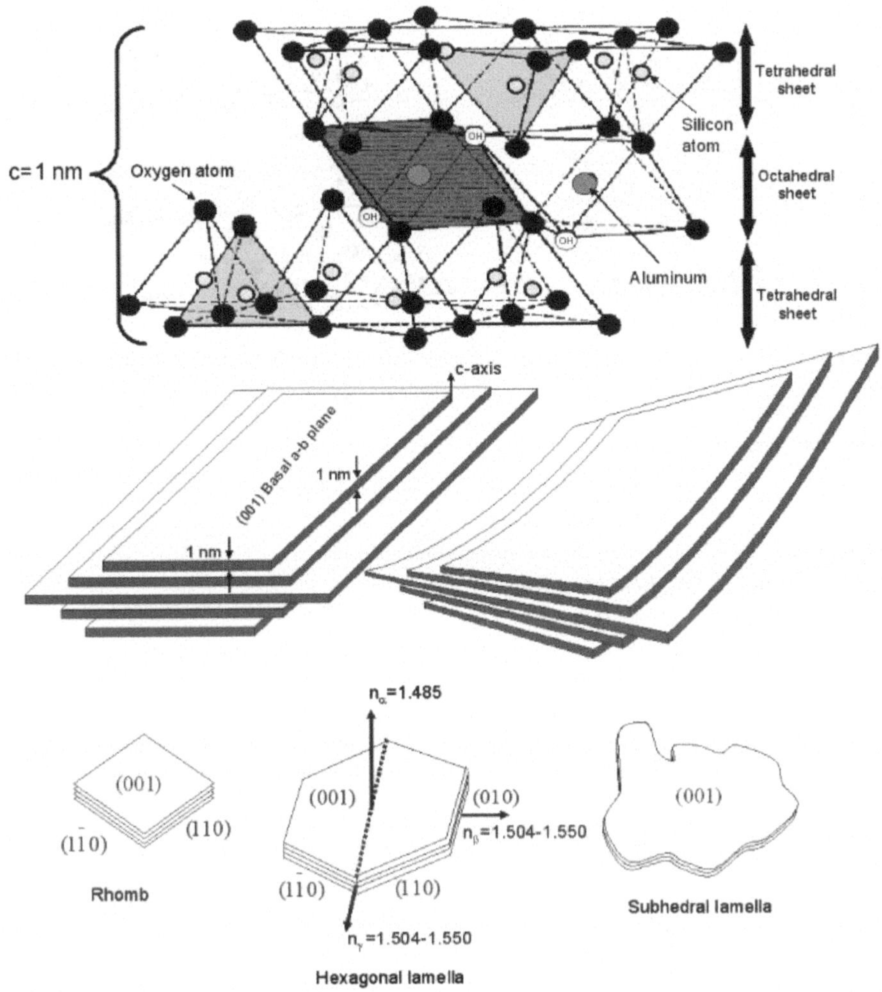

Fig. 3.1 Structure of montmorillonite basic crystal unit and its morphological variations from a perfect hexagonal habit. Reproduced with permission from [5]. Copyright 2004, Elsevier Science Ltd.

The physical mixture of a polymer and LS may not form a nanocomposite. This situation is analogous to immiscible polymer blends, and in most cases separation into discrete phases takes place. In immiscible systems, which typically correspond to the more conventionally filled polymers, the poor physical interaction between the organic and the inorganic components leads to poor mechanical and thermal properties. In contrast, strong interactions between the polymer and the LS in PNCs lead to the organic and inorganic phases being dispersed at the nanometer level. As a result, nanocomposites exhibit unique properties not shared by their micro counterparts or conventionally filled polymers [6–12].

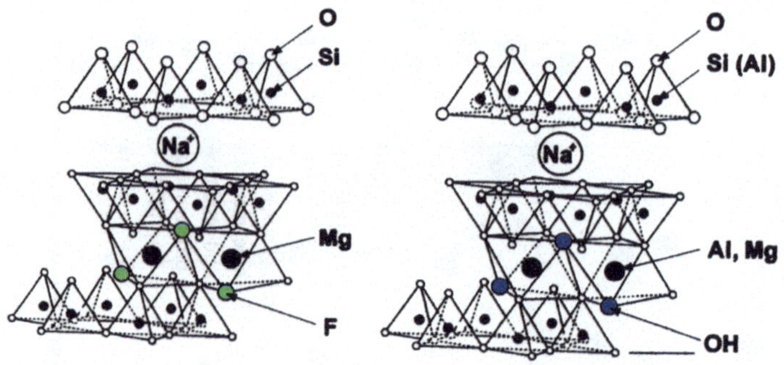

Fig. 3.2 **a** Unit crystal structure of montmorillonite (left side) and **b** synthetic fluorine mica (right side)

Table 3.4 Physical characteristics of mica type clay (muscovite type)

Color	Usually white, silver, yellow, green and brown
Luster	Vitreous to pearly
Transparency	Crystals are transparent to translucent
Crystal system	Monoclinic; 2/m
Unit cell molecular weight	540.5 (g/ml)
Crystal habits	Usually tabular crystals with a prominent pinacoid termination
Field indicators	Crystal habit, cleavage, elastic sheets, color and associations
Cleavage	Perfect in one direction producing thin sheets or flakes. Cleavege sheets are flexible and elastic, meaning they can be bent and will flex back to original shape
Hardness	2–2.5 (in Moh's scale and at room temperature)
Average specific gravity	2.8 g/ml
Facture	Not readily observed due to cleavage but is uneven
Swelling behavior	Very low
Notable occurrences	India, Pakistan, Brazil and many USA localities
Associate minerals	Include quartz, feldspars, beryl and tourmalines

Source Amethyst Galleries, Inc. web sites. © 2006 Amethyst Galleries, Inc.

3.1.2 Structure and Properties of Organically Modified LS (OMLSs)

Purified pristine layered silicates usually contain hydrated Na^+ or K^+ ions [13]. Obviously, in this pristine state, LSs are only miscible with hydrophilic polymers, such as poly(ethylene oxide) (PEO) [14], or poly(vinyl alcohol) (PVA) [15]. To render LSs miscible with other polymer matrices, one must convert the normally hydrophilic silicate surface to an organophilic one, making the intercalation of many engineering polymers possible. Generally, this can be done by ion-exchange reactions with cationic surfactants including primary, secondary, tertiary, and quaternary alkyl ammonium or alkylphosphonium cations. MMT containing Na^+ or K^+ is dispersed in water; its silicate layers swell uniformly and extent of swelling is about 30%. Now if an alkyl-ammonium or—phosphonium salt is added to that aqueous dispersion, the surfactant ions are exchanged with the intergallery cations. As a result of this exchange reaction organophilic LS forms, in which alkyl-ammonium or—phosphonium cations are intercalated between the layers and intergallery height increases. Figure 3.3 represents a schematic view of ion-exchange reaction [16]. By changing the length, type of alkyl chain or by addition of some polar groups, the hydrophilic/hydrophobic and other characteristics of the layered silicate can be adjusted such that surface modification of the clay surface possible.

A representative commercial method for the preparation of orgnophilic MMT (CEC = 110 meq/100 g) by using octadecyl ammonium chloride is described here. The alkyl ammonium cation (it may be primary, secondary, quaternary or tertiary) is dissolved in a 50:50 mixture of ethanol and deionized H_2O at 50–70 °C. In the case of primary amines an equivalent amount of HCI is generally added to the solution. An 1 wt% aqueous suspension of the LS is added to the alkylammonium solution and the mixture is stirred for 5–6 h at 50–70 °C. The cation-exchanged silicates are collected by filtration and subsequently washed with a mixture of hot ethanol and deionized H_2O until an $AgNO_3$ test indicated the absence of halide anions. The filter cake is dried at room temperature, ground, and further dried at 70–80 °C under vacuum for at least 24 h [17].

Table 3.5 summarizes various types commercially available organically modified layered silicates (OMLS) generally used for the preparation of polymer

Fig. 3.3 Schematic picture of an ion-exchange reaction. The inorganic, relatively small (sodium) ions are exchanged against more voluminous organic onium cations. This ion-exchange reaction has two consequences: firstly, the gap between the single sheets is widened, enabling polymer chains to move in between them and secondly, the surface properties of each single sheet are changed from being hydrophilic to hydrophobic. Reproduced with permission from [16]. Copyright 2003, Elsevier Science Ltd.

Table 3.5 Physical properties of commercially available organically modified layered silicates

Commerical name	Organic modifier	Modifier concentration (meq/100 gm)	$d_{(001)}$ spacing (nm)	Weight loss on ignition (%)
From Southern Clay Products, Inc., USA				
Cloisite® Na (CNa)	None	CEC = 92.6	1.17	7
Cloisite® 30B (C30B)	Methyl tallow bis-2-hydroxyethyl quaternary ammonium	90	1.85	30
Cloisite® 10A (C10A)	Dimethyl benzyl hydrogenated tallow quaternary ammonium	125	1.92	39
Cloisite® 25A (C25A)	Dimethy hydrogenatedtallow 2-ethylhexyl quaternary ammonium	95	1.86	34
Cloisite® 93A (C93A)	Methyl dihydrogenatedtallo ammonium	90	2.36	40
Cloisite® 20A (C20A)	Dimethyl dihydrogenatedtallo, quaternary ammonium	95	2.42	38
Cloisite® 15A (C15A)	Dimethyl dihydrogenatedtallo quaternary ammonium	125	3.15	43
Cloisite® 6A (C6A)	Dimethyl dihydrogenatedtallo quaternary ammonium	140	3.59	47

Commercial name	Organic modifier	d_{001} spacing (nm)
From CO-OP Chemical Co., Ltd., Japan		
Somasif™ (ME-100) (CEC = 120 meq/100 g)	None	1.25
MAE	Dimethyl dialkyl (tallow) ammonium	3
MTE	Trioctyl methyl ammonium	2.5
MEE	Dipolyoxy ethylene alkyl (COCO) methyl ammonium	2.3
MPE	Polyoxy propylene methyl diethyl ammonium	5

Commercial name	Organic modifier	Modifier concentration (wt%)	Applicable matrix
From Nanocor Inc., USA			
MMT	None	CEC = 145 meq/ 100 g)	none
I.24TL	12-amino dodecyl acid	–	Polyamide-6 polymerization
I.28E	Tri-methyl stearyl-ammonium	25–30	Epoxy, urethane
I.30E	Octadecyl (stearyl) ammonium	25–30	Epoxy, urethane

(continued)

Table 3.5 (continued)

Commercial name	Organic modifier	Modifier concentration (wt%)	Applicable matrix
Rheospan AS	Di-methyl di-hydrogenated tallow ammonium		Unsaturated polyesters, vinyl esters
I.34TCN	Methyl Octadecyl di-2-hydroxy ethyl ammonium		Polyamide, poly(butylene terephthlate), melt compounding
I.30P	Octadecyl (stearyl) ammonium	25–30	For polyolefin concentrates
I.44PA	Di-methyl di-hydrogenated tallow ammonium		For polyolefin concentrates

nanocomposites. In Table 3.6 we presented chemical structure and abbreviation of most commonly used surfactants.

Alkylammonium or alkylphosphonium cations (see Table 3.6) in the organosilicates lower the surface energy of the inorganic host and improve the wetting characteristics of the polymer matrix, and result in a larger interlayer spacing. Additionally, the alkylammonium or alkylphosphonium cations can provide functional groups that can react with the polymer matrix, or in some cases initiate the polymerization of monomers to improve the strength of the interface between the inorganic and the polymer matrix [13].

Traditional structural characterization to determine the orientation and arrangement of the alkyl chain was performed by Lagaly in 1986 using X-ray diffraction (XRD) [18]. Depending on the packing density, temperature and alkyl chain length, the chains were thought to lie either parallel to the silicate layers forming mono or bilayers, or radiate away from the silicate layers forming mono or bimolecular arrangements [2, 17, 19–21]. Such idealized structures, based almost exclusively on all-trans segments are potentially misleading, since they fail to convey the most significant structural characteristics of aliphatic chains-the capacity to assume as enormous array of configurations because of the relatively small energy difference between trans and gauche conformers (0.6 kcal/mol, 2.5 kJ/mol) [22]. Vaia et al. [17] proposed an alternative arrangement based on a disordered chain configuration containing numerous gauche conformers and this arrangement is consistent with the observed gallery height. These arrangements, however, indistinguishable by XRD, lead to much different interlayer structure and molecular environment.

To understand the layering behavior and structure of confined quaternary alkylammoniums into the two dimensional layered silicate galleries, Zeng et al. [23] performed isothermal-isobaric (NPT) molecular dynamics simulation. Their work was focused on systems consisting of two silicate layers and a number of alkylammoniums, and involves the use of modified Dreiding force field. Figure 3.4 represents a snapshot of the simulation cell of a model octadecyl dihydroxyl ethyl

Table 3.6 Chemical structure of most commonly used of surfactants for the modification of layered silicates

Surfactants	Chemical formula	Abbreviations
Methyl tallow bis-2-hydroxyethyl quaternary ammonium	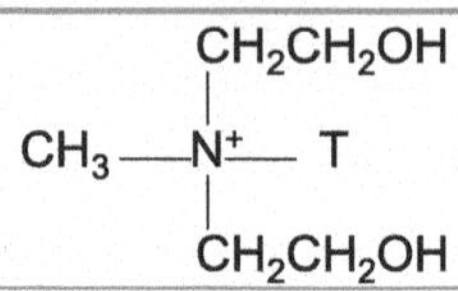	MT2EtOH
Dimethyl dihydrogenatedtallo quaternary ammonium	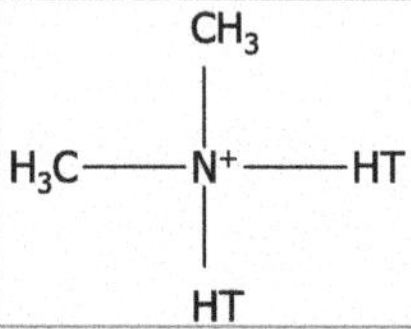	2M2HT
Dimethy hydrogenatedtallow 2-ethylhexyl quaternary ammonium	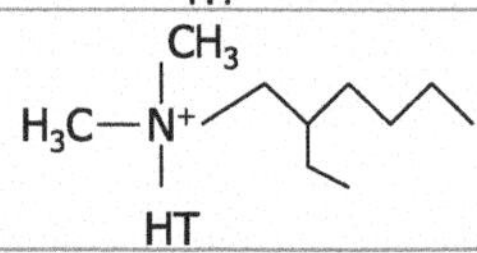	2MHTL8
Dimethyl benzyl hydrogenated tallow quaternary ammonium	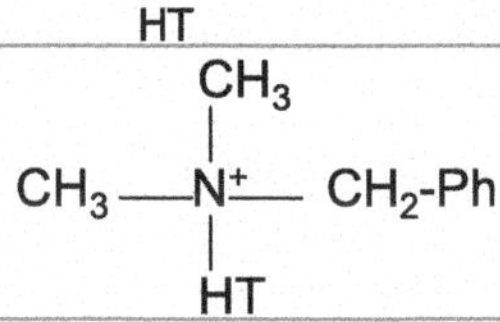	2MBHT
Dimethyl dialkyl (tallow) ammonium	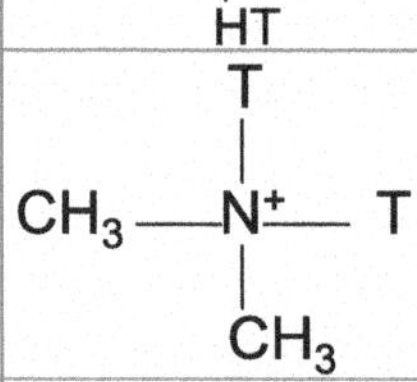	2M2T
Trioctyl methyl ammonium	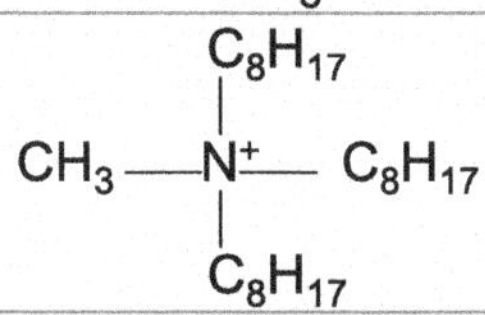	3OM
Dipolyoxy ethylene alkyl (COCO) methyl ammonium	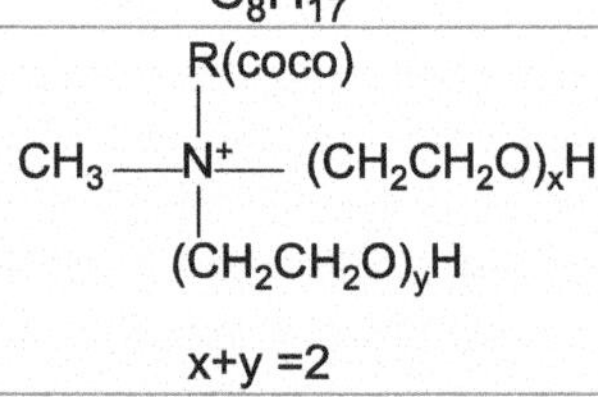	
Polyoxy propylene methyl diethyl ammonium	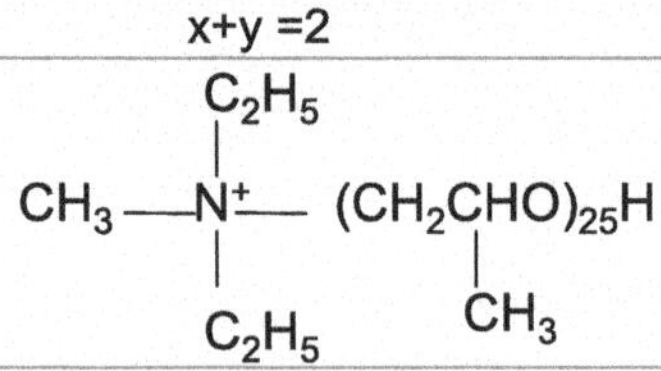	
Octadecyl amine	$CH_3(CH_2)_{16}CH_2NH_2$	ODA

(continued)

Table 3.6 (continued)

Surfactants	Chemical formula	Abbreviations
Dimethyl octadecyl amine	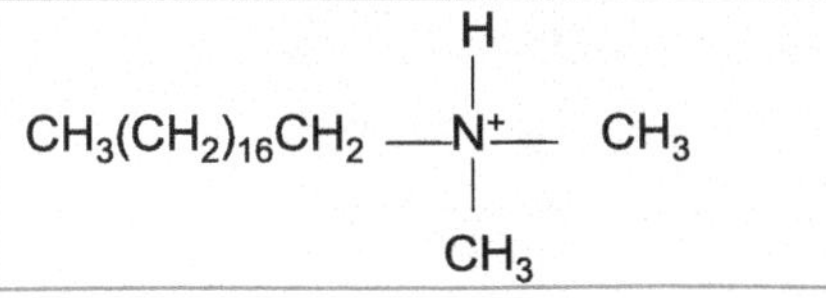	2MODA
Hexadecyl trimethyl ammonium	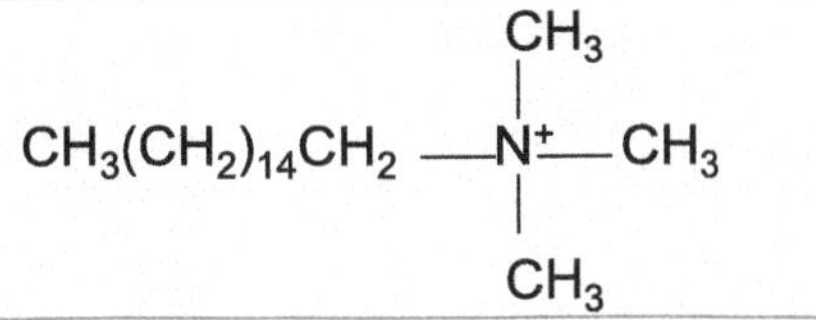	
Dodecyl Triphnyl phosphonium	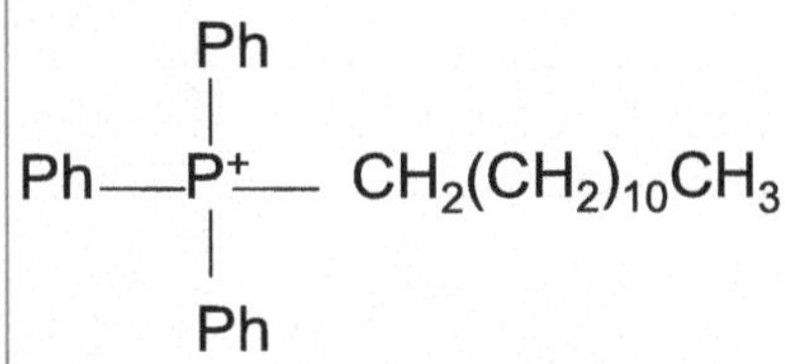	3PDDP
Hexadecyl tributyl phosphonium	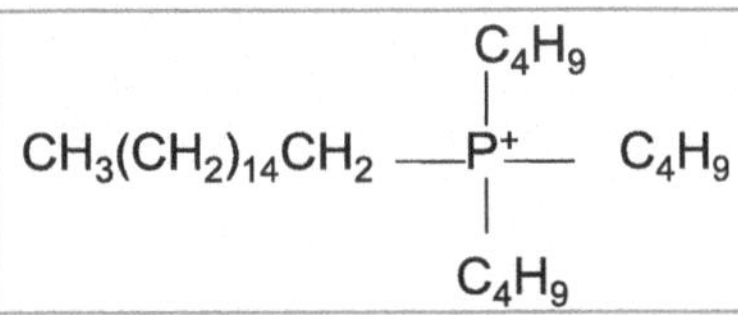	
Dodecyl trimethyl phosphonium	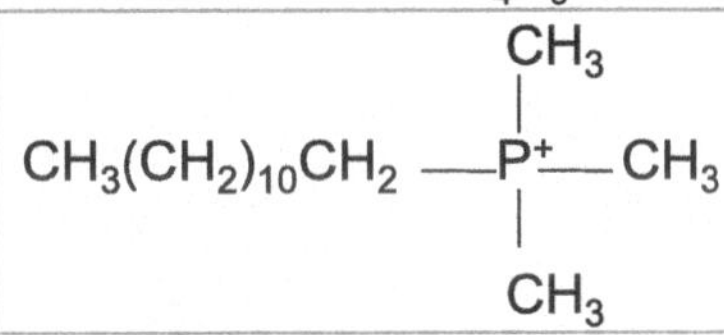	

methyl (DODDMA) modified MMT with CEC of 85 meq/100 g following 800 ps of isothermal–isobaric (NTP) simulation at 300 K. The side view in Fig. 3.4 clearly demonstrates the layering behavior of the surfactant chains within the interlayer space of the MMT. Pseudo-quadrilayer structure is observed and the alkyl chains in each layer adopt an orientation with their longest axis approximately parallel to the MMT surface. The positive charge head groups of surfactant chains are found close to the MMT layers. The methyl carbon atoms observed in the middle layers are mainly attributed to the tail methyl's in the long alkyl chains. It can be seen that all-trans conformation is hard to be realized from their simulated results. Thus, the idealized structural models, such as pseudo-trilayers and paraffin-type monolayers and bilayers, do not directly reveal the significant structural characteristics of alkyl chains.

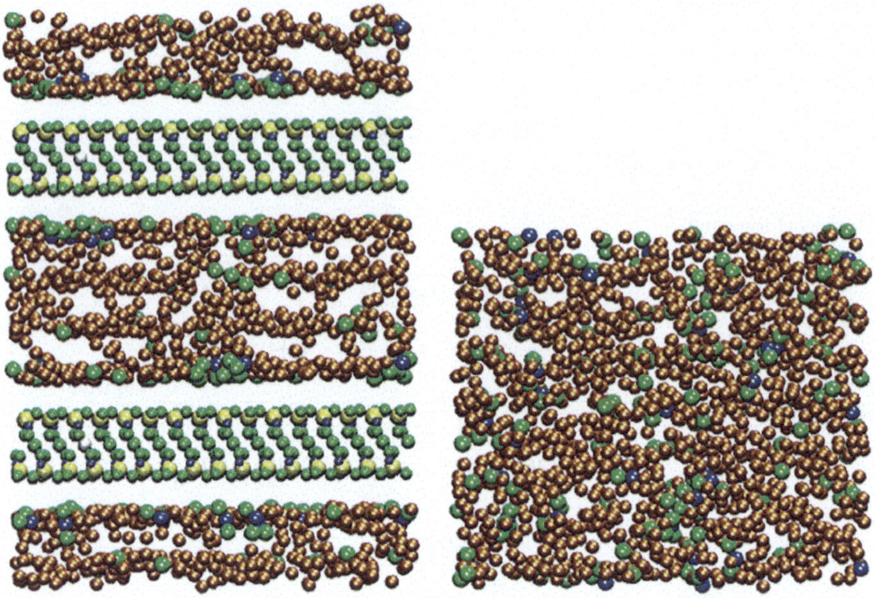

Fig. 3.4 Snapshot at 800 ps for a system at 300 k consisting of two montmorillonite layers and forty dioctadecyl dimethyl ammonium chains viewed from side (left) and normal to the layer surface (right): blue ball (N), green ball (methyl C), organe ball (methylene C). Reproduced with permission from [23]. Copyright 2003, the American Chemical Society

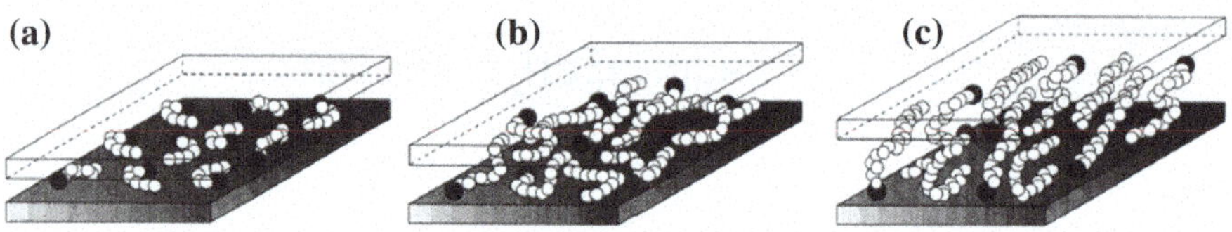

Fig. 3.5 Alkyl chain aggregation models for FH-C. As the number of carbon atoms increases, the chains adopt **n** more ordered structure. For the shortest lengths (**a**), the molecules are effectively isolated from each other. At intermediate lengths (**b**), quasidiscrete layers form with various degrees of inplane disorder and interdigitation between the layers. For longer lengths (e), interlayer order increases leading to a LC environment. Open circles represent CH2 segments while cationic head groups are represented by filled circles. The top silicate layer has been left transparent to improve the perspective of the interlayer. Reproduced with permission from [17]. Copyright 194, the American Chemical Society

These simulated results are, however, consistent with study by Vaia et al. [17] using FTIR techniques. They showed that alkyl chains can vary from liquid-like to solid-like, with the liquid-like structure dominating as the interlayer density or chain length decreases (see Fig. 3.5), or as the temperature increases. This occurs because of the relatively small energy differences between the *trans* and *gauche* conformers; the idealized models described earlier assume all trans conformations. In addition, for longer chain length surfactants, the surfactants in the layered silicate can show thermal transition akin to melting or liquid-crystalline to liquid-like transitions upon heating. In addition, an NMR study reported by Wang et al. [24] indicated the coexistence of ordered *trans* and disordered gauche conformations.

Li and Ishida [25] documented the detail structure of surfactant inside the silicate galleries. They studied the intercalation processes of hexadecylamine into betonite clay and the fine structure of surfactant in nanoscale confined space by differential scanning calorimetry (DSC). A strong layering behavior with an ordered amine arrangement was observed. Less than 35% of the confined amine forms an ordered structure within the silicate galleries and exhibits much higher melting temperature than the free amine. The authors also found that the structure of the confined amine directly related to the initial *d*-spacing of the LS and the amount of amine and the confined amine chains readily nucleate because of their restricted mobility. The ordered confined structured of amine chains would be further influenced by the further intercalation of polymer chains. Further, an important fact is that more than 60% of the nanoscale confined does not exhibit the melting transition.

3.2 Structures of LS-Containing PNCs

The performance of PNCs is improved by merely taking advantage of the nature and properties of the nanofiller [26–28]. When good dispersion of a layered silicate is achieved, the properties of the PNCs are considerably enhanced compared to those of the neat polymer [26]. The dispersion of the silicate particles depends on the interfacial interaction between a polymer matrix, the modified or unmodified layered silicates, processing conditions, and processing strategies [26]. Two types of PNCs such as intercalated and exfoliated are thermodynamically attainable, as demonstrated in Fig. 3.6.

Intercalated PNCs. In this type of PNCs, the inclusion of polymer chains into the two-dimensional silicate galleries happens in a crystallographically steady mode irrespective of the nanoclay to polymer ratio.

Exfoliated PNCs. In the exfoliated PNCs, the individual silicate layers are dispersed and distributed homogeneously in the polymer matrix by the average distance that depends on the nanoclay loading. Generally, the nanoclay content of exfoliated nanocomposites is much lower than that of intercalated PNCs.

Nanocomposites develop or obtain improved properties with a low loading of LS (3–5 wt%), unlike conventional composites that require high content of the filler (10–50 wt%) for practically the same enhanced properties [26].

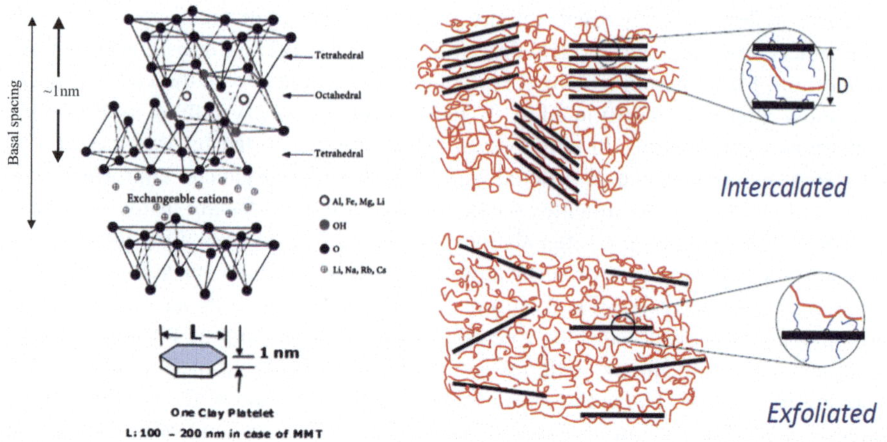

Fig. 3.6 Single crystal structure of MMT and schematic illustration of two different types of thermodynamically achievable layered silicate-containing polymer nanocomposites

3.3 Formation of LS-Containing PNCs

Layered silicate-based PNCs are manufactured mainly using three different methods, such as solution intercalation of the polymer, in situ polymerization, and melt compounding.

3.3.1 Solution Intercalation Method

This method involves using solvents such as water, toluene, and chloroform that swell the silicate layers and can dissolve the polymer. It entails swelling the nanoclay in the solvent, and then mixing with a polymer solution based on the solvent. The mixture is then cast to evaporate the solvent and recover the nanocomposites. The resultant nanocomposites have structures that are intercalated or exfoliated depending on the degree of diffusion of the polymer chains into the nanoclay galleries [26]. A schematic illustration of the intercalation of a polymer from solution is shown in Fig. 3.7a.

The solution intercalation method is based on a solvent system in which the polymer or pre-polymer is soluble and the silicate layers are swellable. The LS is first swollen in a solvent, such as water, chloroform, or toluene. When the polymer and LS solutions are mixed, the polymer chains intercalate and displace the solvent within the interlayer of the silicate. Upon solvent removal, the intercalated structure remains, resulting in PNC.

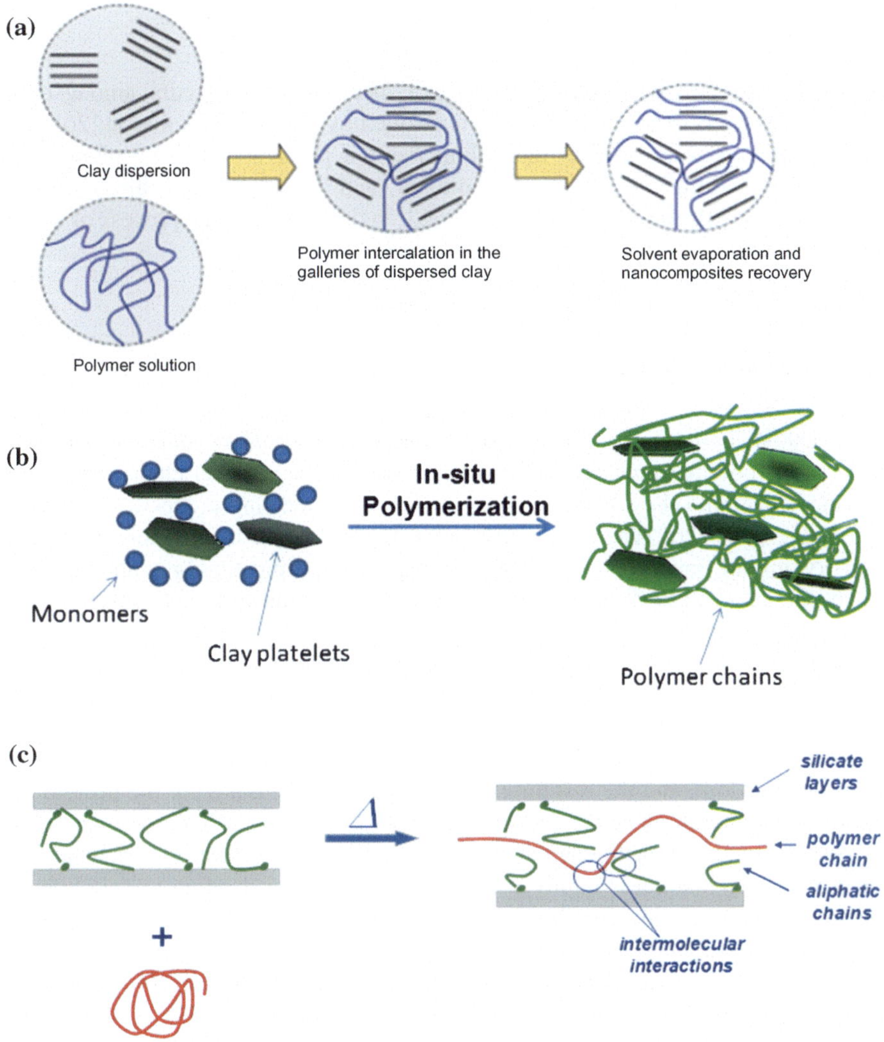

Fig. 3.7 **a** Schematic representation of the intercalation of the polymer from solution, **b** schematic representation of in situ polymerization, and **c** schematic representation of melt intercalation

For the overall process, in which polymer is exchanged with the previously intercalated solvent in the gallery, a negative variation in the Gibbs free energy is required. The driving force for the polymer intercalation into LS from solution is the entropy gained by desorption of solvent molecules, which compensates for the decreased entropy of the confined, intercalated chains. Using this method, intercalation only occurs for certain polymer/solvent pairs. This method is good for the intercalation of polymers with little or no polarity into layered structures, and facilitates production of thin films with polymer-oriented intercalated silicate layers.

3.3.2 In Situ Polymerization

In the in situ polymerization method, polymer formation occurs amongst the intercalated sheets of the nanoclay that is swollen by a monomer solution [26]. The swelling stage occurs prior to the polymerization process, which is initiated via heat, radiation diffusion of a suitable initiator, and a catalyst fixed through cation exchange inside the interlayers [26]. Figure 3.7b demonstrates the process of in situ polymerization.

3.3.3 Melt Intercalation

This method includes a polymer and LS mixture annealed beyond the softening point of the polymer under shear. The dispersion of the polymer chains from bulk polymer melts into the silicate galleries during the annealing process; see illustration in Fig. 3.7c [26]. The melt intercalation method is the most viable method in industry for the preparation of PNCs owing to its compatibility with industrial processes (e.g. extrusion and injection molding), absence of solvent, and environment-friendliness [26].

In recent years, the melt intercalation technique has become the standard for the preparation of PNCs. During polymer intercalation from solution, a relatively large number of solvent molecules have to be desorbed from the host to accommodate the incoming polymer chains. The desorbed solvent molecules gain one translational degree of freedom, and the resulting entropic gain compensates for the decrease in conformational entropy of the confined polymer chains. Therefore, there are many advantages to direct melt intercalation over solution intercalation. For example, direct melt intercalation is highly specific for the polymer, leading to new hybrids that were previously inaccessible.

So far, experimental results indicate that the outcome of polymer chains intercalation into two dimensional silicate galleries depends critically on LS surface functionalization and constituent interactions. Present author observes that (a) an optimal interlayer structure on the OMLS, with respect to the number per unit area and size of surfactant chains, is most favorable for nanocomposite formation, and (b) polymer intercalation depends on the existence of polar interactions between the OMLS and the polymer matrix.

To understand the thermodynamic issue associated with the nanocomposite formation during melt-intercalation, Vaia et al. [29, 30] applied a mean-field statistical lattice model, reporting that calculations based on the mean field theory agree well with experimental results. Although there is entropy losses associated with the confinement of a polymer melt during nanocomposite formation, this process is allowed because there is an entropy gain associated with the layer separation, resulting in a net entropy change near to zero. Thus, from the theoretical model, the outcome of nanocomposite formation via polymer melt intercalation depends primarily on energetic factors, which may be determined from the surface energies of the polymer and OMLS.

Based on the Vaia et al. [29, 30] study and the construction of product maps, general guidelines may be established for selecting potentially compatible polymer/OMLS systems. Initially, the interlayer structure of the OMLS should be optimized in order to maximize the configurational freedom of the functionalize chains upon layer separation, and to maximize potential interaction sites at the interlayer surface. For these systems, the optimal structures exhibit a slightly more extensive chain arrangement than with a pseudo-bilayer. Polymers containing polar groups capable of associative interactions, such as Lewis-acid/base interactions or hydrogen bonding, lead to intercalation. The greater the polarizability or hydrophilicity of the polymer, the shorter the functional groups in the OMLS should be in order to minimize unfavorable interactions between the aliphatic chains and the polymer.

One of the great advantages of the Vaia mean-field statistical lattice model is the ability to determine analytically the effect of various aspects of the nanocomposite formation. According to this model, the variation of the free energy of mixing and subsequent dependence on enthalpic and entropic factors, suggest the formation of three possible structures—phase separated, intercalated and exfoliated. Although, Vaia model able to address some of the fundamental and qualitative thermodynamic issues associated with the nanocomposite formation, however, some of the assumptions such as the separation of configurational terms and intermolecular interaction and the further separation of the entropic behavior of the constituents, somewhat limit the usefulness of the model. Not only that, this model is based on nanocomposites where polymer chains are completely tethered with the silicate surface, which is not the case for most of the PNCs.

To overcome the limitation of Vaia model, Balazs et al. [31] proposed a model based on a self-consistent field (SCF) calculation, such as the Scheutjens and Fleer theory [32]. In the Scheutjens and Fleer theory, the phase behavior of polymer systems is modeled by combining Markov chain statistics with a mean field approximation. These calculations involve a planar lattice where lattice spacing represents the length of a statistical segment within a polymer chain. Details regarding this theory can be found in [32]. Using this method Balazs et al. [31] tried to calculate the interactions between two surfactant-coated surfaces and a polymer melt. They considered two planar surfaces that lie parallel to each other in the *xy* plane and investigated the effect of increasing the separation between the surfaces in the *z* direction. The two surfaces are effectively immersed within a polymer melt. As the separation between the surfaces is increased, polymer from the surrounding bath penetrates the gap between these walls. Each surface is covered with monodispersed end-grafted chains, i.e., surfactants. Now if χ_{surf} represents the Flory–Huggins interaction parameter between the polymers and the underlying solid substrate and χ_{ssurf} represents the Flory–Huggins interaction between the surfactant and surface. Therefore, $\chi_{surf} - \chi_{ssurf} = 0$. It should be noted here; in their calculation Balazs did not consider electrostatic interaction.

Their calculations show that on increasing the attraction between the polymers and the modified surfaces are qualitatively similar to observation as made by Vaia et al. [29, 30]. However, Balazs et al. [31] found that the actual phase behavior and morphology of the mixer can be affected by the kinetics of the polymers penetrate the gap between the plates. At the beginning, the polymer chains have to penetrate the space between the silicate layers from an outer edge and then diffuse toward the centre of the gallery. Now if we consider the case where $\chi_{surf} < 0$ and thus, the polymer and surface experience an attractive interaction. In this case as the polymer diffuses through the energetically favorable gallery, it maximizes contact with the two confining layers. As a results, the polymer 'glues' the two surfaces together as it moves through the interlayer. This 'fused' condition could represent a kinetically trapped state and consequently, increasing the attraction between the polymer and clay sheets would only lead to intercalated, rather than exfoliated structures. On the other hand, in the case where $\chi_{surf} > 0$, the polymer can separate the sheets, as the chain tries to retain its coil-like conformation and gain entropy. However, recent literatures revealed that the melt mixing of OMLSs and almost no attractive polymer matrices always lead to the formation of phase separated structure. The SCF calculations and phase diagrams lead to the same conclusion.

To overcome this problem, Balazs et al. [33] proposed the scheme of using a mixture of functionalized and non-functionalized polymer for the melts. While the stickers at the chin ends are highly attracted to the surface, the remainder of polymer does not react with the substrate. Thus, as the polymer chain penetrate the sheets, the majority of the chain is not likely to glue the surfaces together.

Balazs et al. [34, 35] also proposed a simple model that describes the nematic ordering in the polymer-layered silicate systems. Because of the very high degree of anisotropy (a typical layered silicate platelet is ~1 nm in thickness and 100–200 nm in diameter), layered silicate particles experience strong orientational ordering at low volume fractions and can form liquid crystalline phases such as nematic, smectic, or columnar, in addition to traditional liquid and solid phases. Starting from the Onsager free energy functional for the nematic ordering of rigid rods, the developed a modified expression to combine the disk orientational and positional entropy, steric excluded-volume effects, translational entropy of the polymer, and finally the Flory–Huggins enthalpic interaction. The resulting isothropic-nematic phase diagram correctly represents many important features, such as the role of shape anisotropy in depressing the ordering transition and the increase in the size of the immiscibility region with increases in the polymer chain length. Unlike most of the phenomenological theories of polymer-liquid crystal systems [36–39], in the Onsager-type model the features of the phase diagram are directly derived from the geometric characteristics of the anisotropic component.

Balazs et al. first modified and expanded Onsager theory by including nematic, smectic and columnar crystalline phases. To do calculation, they also adopted Somoza–Tarazona formalism of density functional theory (DFT) and then incorporated expressions that describe the entropy of mixing between the different components and the enthalphic interaction between the platelets [40, 41]. The resulting free energy function can be minimized with respect to both the orientational and positional single-particle distribution function of the platelets, and thus, potentially, all phases and coexistence regions can be determined. The resulting phase diagram was shown to exhibit a strong dependence on the shape anisotropy of the layered silicate particles, the polymer chain length, and the strength of the interparticle interaction. In particular, an increase in the shape anisotropy for oblate ellipsoidal filler particles leads to the broadening of the nematic phase at the expense of the isotropic region. The increase in the polymer chain length leads to the formation of the crystalline and/or liquid crystalline mesophases and promotes segregation between polymer-rich regions and filler particles. Finally, an increase in the strength of the interparticle potential leads to the complete elimination of the nematic phase and to the direct coexistence between isotropic and crystal or columnar phases. The only limitation of this model is that this model cannot determine the topology of the phase diagram and the nature of the ordered phases for intermediate and high volume fractions of colloidal particles.

The huge interfacial area and the nanoscopic dimensions between nanoelements differentiate PNCs from conventional composites and filled plastics. The dominance of interfacial regions resulting from the nanoscopic phase dimensions implies that the behavior of PNCs cannot be understood by simple scaling arguments that begin with the behavior of conventional polymer composites. Since an interface limits the number of conformations that polymer molecules can adopt, the free energy of the polymer at the interfacial region is fundamentally different from that of polymer molecules far away from the interface, i.e., bulk. The influence of interface always depends on the fundamental length scale of the adjacent matrix. In the case of polymer molecules, this is of the order of the radius of gyration of a polymer chain, R_g and this is equal to 5–10 nm. For this reason, in nanocomposites with a very few volume percent of dispersed nanofillers in polymer matrix, the entire matrix may be considered to be a nanoscopically confined interfacial polymer. The restrictions in chain conformations will alter molecular mobility, relaxation behavior, free volume, and thermal transition such as the glass transition temperature.

Recent molecular dynamics simulation studies have shown that the dynamics of the polymer chains undergo radical changes at the interfacial region. For example, both the chain mobility and the chain relaxation times can be slowed by three orders of magnitude near physisorbing surfaces. Not only that, extensive surface forces apparatus experiments report how these novel dynamics of nanoconfined polymers are manifested through viscosity increases, two values orders of magnitude higher than the bulk values solid-like responses to imposed shear, and confinement induced 'sluggish' dynamics that suggest the existence of a 'pinned', 'immobilized' layer adjacent to the confining mica surface.

The kinetics of intercalation of polymer chains into the silicate galleries to form layered nanocomposites has been studied by Vaia et al. [30]. They have investigated the kinetics of the intercalation of PS above its entanglement molecular weight into octadecylammonium exchanged fluorohectorite. XRD reveals the average layer spacing in the unintercalated silicate to be 2.13 nm. During intercalation, this spacing increases to 3.13 nm. Measurement of this layer spacing as a function of time during intercalation yields a time-dependent fraction of intercalated silicate that is directly comparable to the time dependent number of beads in the slit, $\chi(t)$ and it is well fit by the prediction of a continuum diffusion model. In their model Vaia et al. [30] considered the diffusion of polymer chains into an empty cylinder with a permeable wall and impermeable caps. The diffusion coefficient determined by fitting $\chi(t)$ to this model, effective diffusion coefficient, D_{eff}, is large compared to the equilibrium self-diffusion coefficient in bulk polystyrene of the appropriate molecular weight. These results suggested that the process of intercalation of polymer chains into the two dimensional silicate galleries is limited by the transport of the polymer chains into the primary particles of the silicate and not specifically by transport of polymer chains inside the silicate galleries.

Now to understand the detail formation kinetics and physical properties of PNCs we need to have a clear molecular picture of the structure and dynamics of confined polymers. Researchers generally used the surfaces force apparatus [42–46] and computation studies [47–57] to understand the behavior of confined fluids. Confinement of a fluid on length scales comparable to the molecular size has been demonstrated to dramatically alter its structure and dynamical pictures [57]. For example, confined fluids have been shown to solidify or vertify at temperatures well above the bulk transition temperature [55]. Because of the strong interactions between the confined molecules and the atoms or molecules of the confining medium, the mobility of the molecules in the confined environment is greatly reduced compared to the bulk [42, 58–61]. Molecules in films of nanometer thickness organize in layers parallel to the surface. However, the confining medium induces two dimensional orders in these layers. On the other hand, in certain circumstances, confinement phenomenon may also have opposite effect of enhancing molecular nobilities in a supercooled thin film, relative to the bulk.

Lee et al. [62] have presented an investigation of the molecular mechanism of polymer melt intercalation using molecular dynamics simulations. They tried to find out the motion of polymer chains from a bulk melt into a confined volume. In their model they represented macromolecules by bead-spring chains, leave a reservoir of bulk melt to enter a slit of rectangular cross section and fixed dimension. They adopted a coarse-grained description of polymers, because such a picture has been demonstrated to provide a useful description of melt dynamics over longer time scales than would be accessible with an atomistic model [63, 64]. They also considered a slit of fixed dimension to understand the transport of a polymer melt from the bulk into a confined volume of fixed dimension. However, they did not consider the presence of surfactant molecules in the slit and the swelling of the slit during intercalation.

They found that the intercalation process can be approximately characterized by an effective diffusion coefficient that is twice as large as the equilibrium self-consistence in the bulk melt. Increasing the polymer-silicate interactions is found to induce spontaneous intercalation, but for a high-polymer silicate affinity, the amount of intercalated material at a given time is reduced compared to the case of a weaker polymer-silicate attraction. The crossover from polymer-silicate miscibility to intercalated structures with increasing polymer-surface affinity has already been mentioned by Vaia et al. [30]. However, their study suggested that an important role may be played by the relaxations of polymer bridges that connect the two silicate surfaces. The number of these bridges, as well as their dynamical properties, will be controlled by chain length [65].

Manias et al. [66] extended Vaia et al. studies and presented a systematic study of the kinetics of polymers entering 2 nm wide galleries of mica-type silicates as a function of polymer molecular weight and polymer-surface interactions. They varied the polymer-surface interactions in two ways: either through changing the surface modification, i.e. by varying the organic coverage or through attaching strongly interacting sites-sticky groups-along the polymer chain. The polymer-surface affinity is the one of the most crucial parameters during nanocomposite preparation, since it controls the polymer-surface monomeric friction coefficient and thus determines the motion of the polymer next to a solid surface. The interaction between the polymer and silicate surface can be controlled in two novel and well-controlled ways: (i) keeping the polymer the same and modifying the silicate surface, via controlled surface covering by surfactants of varying length at the same grafting density, and (ii) keeping the organically modified surface the same and modifying the polymer friction coefficient, by attaching along the polymer chain a controlled amount of groups that interact strongly with the silicate surface.

Manias et al. [66] used the PS as polymer and the same octadecylammonium modified fluorohectorite to study the intercalation process. While Vaia et al. [30] made all XRD measurements on similar hybrids during in situ annealing; Manias et al. [66] used ex situ sample for XRD and other measurements. The ex situ method has several advantages over the in situ method: First, one can able to anneal the sample under vacuum, thus reducing any polymer chain degradation. Second, one can heat and subsequently investigate the same side/surface of the pellet sample by XRD, a procedure that provides a much more accurate control of the annealing temperature than in situ case, where pellet is heated in one side and is studied by XRD on the opposite side. They made several comparative studies by using concurrent in situ small-angle neutron scattering (SANS) and intermediate-angle neutron scattering (IANS) to monitor the changes in dimensions of the polymer, i.e., R_g, and also to follow the changes in the single chain scattering function during intercalation.

Results of concurrent SANS and IANS studies shown that during intercalation the silicate gallery expansion directly reflects the motion inside the 2 nm slit pore of the polymer chains. For the same polymer and the same annealing temperature, they found the experimentally measured effective diffusion coefficient depends strongly on the surfactant used for the modification of fluorohectorite. For several different polymer molecular weights and annealing temperatures they observed that D_{eff} increases markedly with longer surfactant lengths and much more than is expected just from the enhancement of polymer mobility resulting from the polymer dilution by small hydrocarbon oligomers. Longer surfactants result in less silicate surface area exposed to the polymer, thus effectively reduced the density attraction sites. On the other hand, introducing controlled amount of groups along the polymer chain that interact strongly with the silicate surface, resulting in a strong decrease of D_{eff}. Therefore, increasing either the density or the strength of these attractive sites leads to much slower intercalation kinetics. However, an increase in site density or strength must also increase the driving force for intercalation; evidently such increases depress the friction coefficient ζ much more strongly.

To understand the atomistic details of the structure of the confined-intercalated-polystyrene chains inside the two dimensional silicate gallery, Manias et al. also carried out molecular dynamic simulation. Details regarding simulation can be found in [66]. They used the rotational-isomeric-state (RIS) model to create initial polymer conformations of PS oligomers. Conformations that fit in the inter-layer gallery were chosen, and the PS chains were equilibrated by an off-lattice Monte Carlo scheme that employed small random displacements of the backbone atoms and orientational biased Monte Carlo rotations of the phenyl rings; at the same time, the surfactants were equilibrated by a configurational biased scheme in coexistence with the polymer chains. After equilibrium, MD simulations were used to obtain the structure and density profiles of the intercalated polymer/surfactant films. The numbers of polymer chains and alkylammonium surfactants were chosen so as to match the densities found in the experimental studies. The results suggested that the confined film adopts a layered structure normal to the solid surfaces, with the polar phenyls dominating the organic materials adsorbed on the walls, and the aliphatic groups predominantly in the center of the pore. ^{1}H–^{29}Si cross-polarized nuclear magnetic resonance (NMR) measurements revealed a coexistence of ultra-fast and solid-like slow segmental dynamics throughout a wide temperature range, below and above T_g, for both the styrene phenyl and the backbone groups. The mobile moieties concentrate at the center of the slit pore, especially for the higher temperatures. This leads to a strong density inhomogeneity in the direction normal to the surface. Fast dynamics occur in the lower density regions, whereas slower dynamics occur in high-segment-density regions close to the surface. This heterogeneous mobility combined with an observed persistence of mobility below the bulk glass transition temperature has implications to nanocomposite properties.

3.4 Ways of Enhancing Dispersion

In recent years, numerous techniques of improving the dispersion of layered silicates in a polymer matrix have been reported in various studies [26]. These techniques include water-assisted extrusion, supercritical carbon dioxide, the use of a compatibilizer, enthalpic methods, and the optimisation of kinetic factors. Lee et al. [67] reported the water-assisted extrusion of polypropylene (PP)/layered silicate nanocomposites at a high shear rate. As revealed by the authors, well dispersed and distributed MMT in PP can be obtained at high shear rates and at 5 wt% MMT loading in water, owing to the swelling of nano-platelets in a slurry state. It is then easier to separate/peel the nanosilicate platelets under shear stress during extrusion. High screw speed can supply enough shear stress. However, average and slow screw speeds were not investigated. The results are only based on one screw element design and water degraded the polymer while inside the extruder barrel.

Nguyen and Baird [68] reported an improved technique for exfoliating and dispersing nanoclay particles into the polymer matrix using supercritical carbon dioxide. The authors established that the existence of CO_2 as a polar organic solvent in a supercritical state showed a good amount of exfoliation using a high content of nanoclay (6.6 wt%). The results were only reported for a screw speed of 15 rpm. The effects of different levels of screw speed have yet to be reported, and this process needs a special extruder to produce the nanocomposites. Deka and Maji [69] reported a study based on the effect of a coupling agent and the nanoclay properties of HDPE, LLDPE, PP, PVC blends, and phargamites karka nanocomposites. It was discovered that improved nanoclay exfoliation can be achieved using a 1 wt% and 3 wt% loading of phargamites (phr) nanoclay and polymer blends. However, this system used different kinds of polymers for blending, which leads to higher costs of the end product. Even after nanoclay surface modification and the addition of a compatibilizer in the polymer matrix, nanoclay agglomerates continued to exist in the polymer matrix.

Another aspect we have to consider the optimization of kinetic factors. In this direction, Li et al. [70] showed that the dispersion of layered silicate was related to the mixing intensity and residence time. Longer residence time ensured better intercalation and exfoliation, while intensive mixing when producing nanocomposites increased microdispersion, exfoliation, and matrix degradation. Furthermore, the samples prepared with the master-batch (MB) method had better mechanical properties than those prepared by single-pass (SP) method, in terms of tensile properties. However, only one screw profile for twin-screw extruder (TSE) was investigated; other profiles were not considered except for the single screw extruder (SSE). Low feeding of the material in the extruder was used, which brings one question in mind: How does the average and high feeding rate affect the dispersion of silicate platelets? However, Lertwimolnun et al. [71] reported that the state of exfoliation was significantly improved when the feed rate decreased. The enhancement of the layered silicate dispersion was assumed to be related to the corresponding increase in the residence time as the feed rate decreased. However, it is well known that longer residence time in the extruder leads to material degradation. Nevertheless, achieving a good level of nanoclay dispersion remains problematic and needs optimal processing conditions in order to expand PNC applications.

Generally speaking, it is well-admitted in the literature that an increase in screw speed leads to a better dispersion [26]. This can be explained by the fact that a higher shear rate allows the breakdown of agglomerates into smaller aggregates. However, a negative effect of the screw speed on exfoliation was also observed in the specific case of a polybutadiene/OMLS system.

The influence of other parameters, including feed rate and temperature, has also been reported in the literature [26]. Increasing feed rate induces mainly a strong reduction in residence time and specific energy. The improvement of exfoliation was assumed to be related to the corresponding increase in residence time as the feed rate decreased (Chap. 5, vol 1). The feed rate and screw speed could be independently varied; materials could be fed or removed at different positions along the screws, etc. Evidently, all of these aspects have an important influence on the microstructure (state of intercalation and exfoliation) of the organoclay.

In a very recent work, Bandyopadhyay et al. [72] tried to find out how the mode of layered silicate addition inclusion affects the morphology and hence the rheological properties of the final composite. In the reported study reported, nylon6/ethyl–vinyl-alcohol (N6/EVOH) was selected to model a blend system, and the effect of the mode of OMMT (organically modified MMT) inclusion on the morphology development and melt-state viscoelastic properties of ternary blend composites was investigated. For the preparation of OMMT-containing N6/EVOH blend composites, both MB and OMMT were processed with an 80N6/20EVOH blend. During processing, the temperatures chosen for different zones of the extruder were 160, 230, 250, 270, 270, and 240 °C, and that for the die was 240 °C. The screw speed used was 400 rpm. A twin-screw extruder (Process 11 co-rotating twin-screw extruder, $L/D = 40$, Thermo Scientific, USA) was used for processing, and extruded samples were collected via a water bath and then pelletized. The authors also prepared MB-containing composites of N6 and EVOH. The silicate content (inorganic part) in all the composites was kept constant at around 3 wt%. Polymer blends without nanoclay, such as N6/EVOH and N6/EVOH/PP-g-MA, were also prepared under the same processing conditions. Table 3.7 summarizes the composition of various samples.

The morphological study using scanning electron microscopy, transmission electron microscopy, three-dimensional tomography, and differential scanning calorimetry indicated that the intercalated silicate layers were located in the interphase region between N6 and EVOH in the N6/EVOH/MB composite and core–shell particles were formed, with EVOH as the core. On the other hand, the intercalated silicate layers were well distributed in the blend matrix of the N6/EVOH/OMMT composite, and it was difficult to differentiate between two phases.

On the basis of experimental evidences, the authors concluded that the mode of nanoclay inclusion could steer the morphology development (Fig. 3.8) and

Table 3.7 Compositions of various blend and composite samples

Samples	Polymer (wt%)[a]	PP-g-MA (wt%)	MMT (%)	Surfactant (wt%)	Antioxidant (irganox B225) (wt%)
MB	0	47.01	30.83	17.53	4.62
N6/EVOH (80:20)	100	0	0	0	0
N6/EVOH/ PP-g-MA	95.42	4.75	0	0	0
N6/EVOH/ OMMT	94.84	0	3	1.71	0.45
N6/MB	90.27	4.57	3	1.71	0.45
EVOH/MB	90.27	4.57	3	1.71	0.45
N6/EVOH/ OMMT	90.27	4.57	3	1.71	0.45

[a]Either N6, or EVOH or N6/EVOH has been considered as polymer in the table. MB, masterbatch; OMMT, organically modified MMT

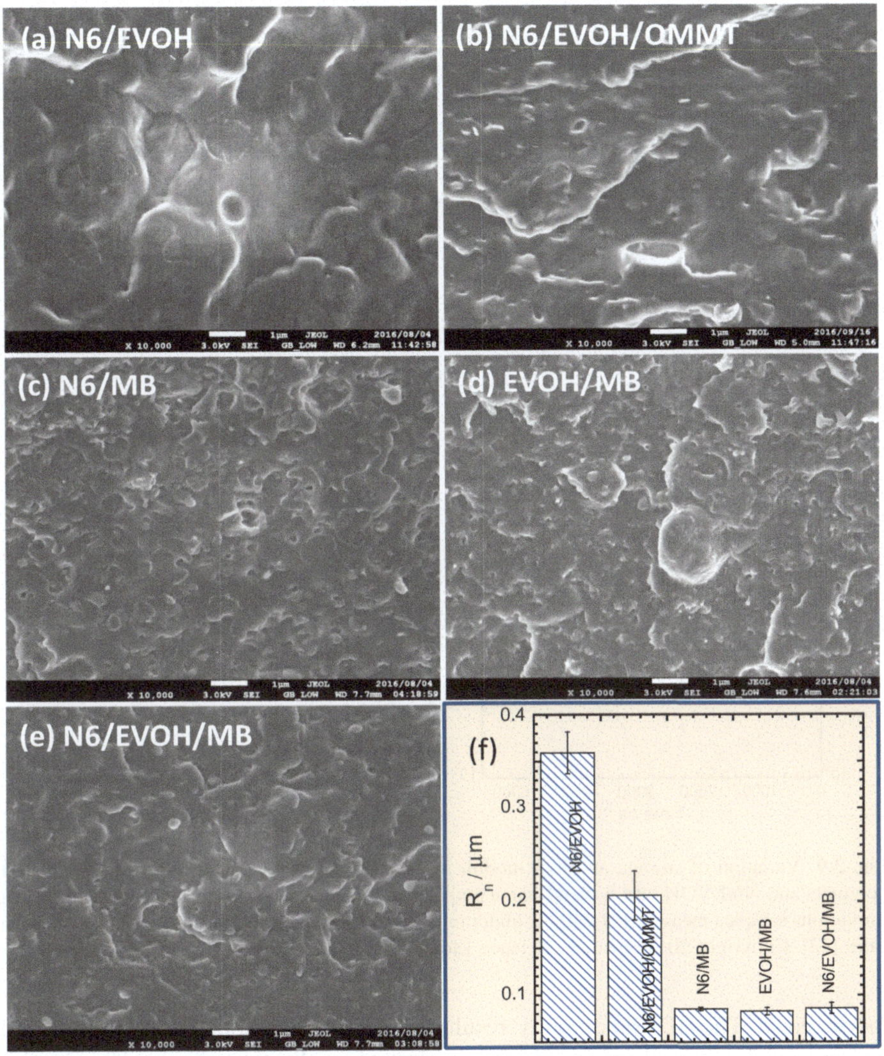

Fig. 3.8 a–e the cryogenically fractured SEM surface images of various blend and blend composites. **f** The number-averaged droplet radii (R_n) was estimated by analyzing 50–100 droplets from several SEM images captured for each sample. Reproduced with permission from [72]. Copyright 2017, Elsevier Science Ltd.

melt-state viscoelastic (Fig. 3.9) properties of the blend composites. The correlation of the structure and morphology with the thermal and rheological properties revealed that the MB played an important role in controlling the interfacial adhesion between N6 and EVOH owing to the reactive compatibilization. Such compatibilization led to the localization of the intercalated silicate layers in the interfacial region which in turn suppressed the coalescence and stabilized the blend

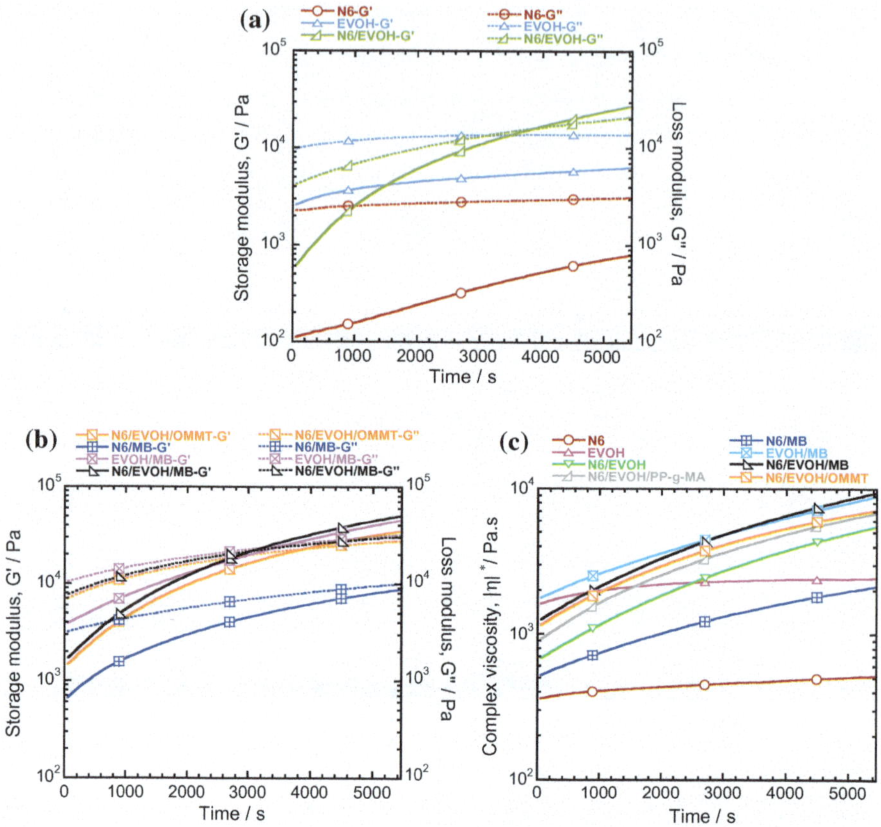

Fig. 3.9 Variation of storage and loss moduli in the melt state as a function of time for **a** neat polymers and N6/EVOH and **b** blend nanocomposites; **c** complex viscosity as a function of time for various samples measured at 240 °C under nitrogen environment. Reproduced with permission from [72]. Copyright 2017, Elsevier Science Ltd.

morphology. The 3D tomography results (details can be found in Chap. 4) and melting and the crystallization behavior confirmed localization of nanoclay particles in the interfacial region and the formation of core–shell particles. The rheological characterizations stipulated that the core–shell blend morphology impeded the gel formation at high rate and temperature. During melt-state polymer processing (e.g., film blowing) the gel or entangled structures caused die swell. Although the die swell could be controlled by designing a longer die or by lengthening the residence time, the consequence of such approaches could lead to the degradation of polymers or the surfactant of the OMMT during processing. Since the N6/EVOH/MB exhibited less gel formation and hence less die swelling when compared with the N6/EVOH/OMMT, the master batch can be considered more effective for the development of nanoclay-containing blend composites.

3.5 Conclusions

During layered silicate-containing polymer nanocomposite formulation, the nano-level dispersion of the silicate particles is the most important characteristic to achieve, in order to have increased interfacial or surface area for polymer-filler interaction, improved co-operative phenomena among dispersed particles, and/or a higher degree of confinement effects. In the case of most polymer matrices, the primary challenge is to find the right chemistry to provide the best thermodynamic driving force to silicate particles at a nano-level. On this issue, researchers are using two approaches. One approach involves the incorporation of functionality into the polymer matrix by grafting, co-polymerization, or blending with other polymers. Another approach is the functionalization of nanoparticle surfaces to improve the compatibility with the polymer matrix. Moreover, in the case of layered silicates that have layered structure, researchers are using ion-exchange chemistry to decrease the inherent van der Waals forces among silicate layers to improve the delamination of silicate platelets in the polymer matrix.

Processing conditions (e.g. temperature profile, feed point, screw speed, feed rate, and screw element configuration) and how nanocomposites are prepared in the extruder have a vital effect on the dispersion of the silicate platelets. The resultant morphology of nanocomposites is not only a question of shear stress or residence time, but also a result of the entire mechanical and thermal history of the material when extruded. The chapter 5 (vol 1) extensively investigate the aspects of processing conditions, such as temperature profile, feed point, screw speed, feed rate, and screw element configuration, and the relationship between the different parameters (optimal conditions). The effects of organically modified layered silicate and compatibilizer loading on the dispersion of silicate layers in the polypropylene nanocomposite have also been investigated.

Acknowledgements The authors would like to thank the Department of Science and Technology and the Council for Scientific and Industrial Research for their financial support.

References

1. Source: Amethyst Galleries, Inc. web site, accessed on April 18, 2006.
2. Brindly SW, Brown G, editors. Crystal structure of clay minerals and their x-ray diffraction. London: Mineralogical Society; 1980.
3. Ray SS, Okamoto K, Okamoto M. Structure-property relationship in biodegradable poly (butylene succinate)/layered silicate nanocomposites. Macromolecules. 2003;36:2355–67.
4. Source: CO-OP Chemical Company, Japan web site, accessed on July 20, 2005.
5. Yalcin B, Cakmak M. The role of plasticizer on the exfoliation and dispersion and fracture behavior of clay particles in PVC matrix: a comprehensive morphological study. Polymer. 2004;45:6623–38.
6. Giannelis EP. Polymer layered silicate nanocomposites. Adv Mater. 1996;8:29–35.

that the results do not always meet expectations. Therefore, elucidating the structure–property relationship of PNCs enables further understanding of the unique, but challenging, aspects of the materials. This chapter provides an overview of the different structural characterization techniques used to analyze PNCs. These methods can be categorized according to the table presented in Fig. 4.1 [1]. The reciprocal space scattering techniques are extremely powerful and enable analysis of the bulk properties of the material. On the other hand, microscopy techniques and tomography provide a direct view of the morphology and degree of the dispersion of the NPs in the polymer matrix. However, care must be taken to interpret the images correctly and avoid artefacts. The interfacial area can be analyzed by electron paramagnetic resonance (EPR), nuclear magnetic resonance (NMR), and optical and dielectric spectroscopy methods. The physical properties can be analyzed via the rheological, mechanical, and barrier performances.

4.2 Reciprocal Space Analysis

4.2.1 Small-Angle X-ray Scattering

The small-angle X-ray scattering (SAXS) method investigates "particles" which are composed of molecules grouped together in a random manner. It is a powerful technique for probing the distribution of NPs, as well as their shape, size, internal structures, crystal lamellar thickness, and surface per volume and/or mass.

4.2.1.1 Dispersion and Distribution of Nanoparticles in PNCs

In the case of PNCs, one of the biggest advantages of SAXS is that it probes materials with sub-nanometer resolution (1–100 nm) while analyzing a large sample size that covers a large number of NPs [2]. SAXS analysis is mainly applied for randomly oriented and statistically distributed particle systems. Hence, their 3D scattering pattern represents the orientational average of their structure. Only in the

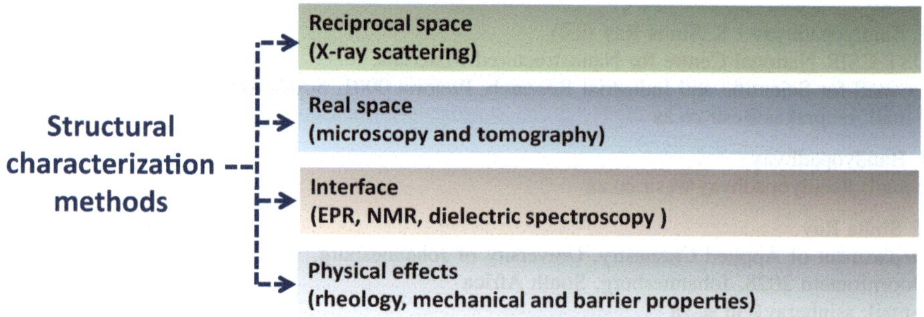

Fig. 4.1 The structural characterization methods used to analyse the polymer nanocomposites

case of three types of ideal symmetry (spherical, cylindrical, and lamellar structures with a centrosymmetric scattering length density distribution) there is no information loss due to orientational averaging [3]. This orientational averaging is radially symmetric and can therefore be reduced to the 1D angle-dependent scattering intensity function $I(q)$ shown in (4.1).

$$I(q) = 4\pi \int_{0}^{\alpha} p(r) \frac{\sin qr}{qr} dr \tag{4.1}$$

where q is the scattering vector and can be related to the scattering angle (θ) and wavelength (λ) by (4.2).

$$q = \frac{4\pi}{\lambda} \sin\theta \tag{4.2}$$

The term $p(r)$ in (4.1) is the pair-distance distribution function of the electrons, which corresponds to the radial or spherical symmetric correlation function of electron density differences weighted by $4\pi r^2$. This term directly shows the probability of finding a pair of electron densities at a particular distance r. All information from the experimental curves in the small angle region is in reciprocal space as $q \propto 1/\lambda$. Therefore, it is difficult to yield direct information about the form and structure of the particles.

If we consider a composite particle consisting of sub-particles with a fixed orientation, the positioning of the centers of mass of the sub-particles can be written as $r_1, r_2, \ldots r_j \ldots r_N$. The scattered wave amplitudes from these sub-particles (with respect to each center) can be designated as $F_1, F_2, \ldots, F_j \ldots, F_N$. The positions of the sub-particles are accounted for by an additional phase factor e^{-iqr_j}. Therefore, the total amplitude of the composite particle can be defined as follows:

$$F(q) = \sum_{1}^{N} F_j(q) \,.\, e^{-iqr_j} \tag{4.3}$$

In general, each amplitude will also have a phase; therefore, F_j can be defined as:

$$F_j = |F_j| \,.\, e^{i\varphi_j} \tag{4.4}$$

Then, intensity ($I(q)$) can be written as:

$$\begin{aligned} I(q) &= FF^* \\ &= \left\langle \sum_{j=k} \sum F_j F_k^* \,.e^{-iq(r_j - r_k)} \right\rangle + 2\left\langle \sum_{j \neq k} \sum |F_j||F_k| \cos(qr_{jk} + q_k - q_j) \right\rangle \end{aligned} \tag{4.5}$$

The double sum contains N terms with $j = k$, where the phase factor consequently vanishes. The remaining term with $j \neq k$ represents the interference between the sub-particles according to the relative distance $r_{jk} = (r_j - r_k)$. Since each pair is counted twice with $r_{jk} = - r_{kj}$, only the real part is considered. Therefore, the intensity contribution with $j = k$ is considered the form factor and that with $j \neq k$ is considered the structure factor.

The real space transformation of the SAXS data (after desmearing) by inverse Fourier transformation (IFT) of the Fredholm integral equation (see (4.1)) can determine parameters such as $p(r)$, from which the form and structure factors can be evaluated. However, in this case, performing an IFT is impossible due to the termination effect of the q-scale and the influence of remaining background scattering, which can cause strong artificial oscillations ("Fourier ripples") in the $p(r)$ function and render the results useless [4, 5]. At small q-values, the measurement is limited by the unscattered primary beam and at large q-values by the progressive decrease in the signal-to-noise ratio. The scattered intensity is usually determined at discrete points. According to counting statistics, the standard deviation of each data point is equal to the square root of the number of pulses registered by the counter. The termination effect can be reduced by extrapolating the scattering curve. For example, the Guinier approximation [4, 5] can be used to extrapolate the scattering curve to a zero angle provided that the first data point is measured at a very small angle. The extrapolation to large angles can sometimes be performed using Porod's law. The termination effect can be minimized by the indirect Fourier transform method developed by Glatter [6, 7]. In most cases, researchers are interested in studying the structure of particles dispersed in solution; to avoid background scattering, the solvent is considered as a background, then the $I(q)$ of the solvent is subtracted from the $I(q)$ of the solution. In the case of polymer nanocomposites, in order to obtain information about the dispersed NPs, the response of the pure polymer should be taken as a background and subtracted from the scattering intensity of the nanocomposite. The generalized indirect Fourier transformation (GIFT) has the following advantages: single-step procedure; optimized general function system; weighted least square approximation; error propagation; minimization of the termination effect; and consideration of the physical smoothing condition given by the maximum intra-particle distance [5]. Therefore, for smoothing conditions it is necessary to estimate the upper limit of the largest particle dimension D_{max}. Therefore, if

$$r \geq D_{max}, \quad p(r) = 0 \tag{4.6}$$

In addition, a function system should be defined in the range $0 \leq r \leq D_{max}$ and a linear combination of these functions should provide $p(r)$. Therefore,

$$p_A(r) = \sum_{\upsilon=1}^{N} c_\upsilon \varphi_\upsilon(r) \tag{4.7}$$

where the suffix "*A*" denotes only that this $p(r)$ is approximated. N is the number of functions and should be chosen to sufficiently cover the range $0 \leq r \leq D_{max}$; c_{υ} contains the unknowns and can be determined by a weighted least square approximation of the experimental data. $\varphi v(r)$ are the cubic *B*-spline functions, which can be defined as multiple convolution products of a step function, representing curves with a minimum second derivative. Each individual spline function can be subjected to a Fourier transform (T_1), wavelength integral (T_2), slit-length integral (T_3), and slit-width integral (T_4). The intermediate result after Fourier transformation of all the splines represents the scattering intensity without the collimation effect corresponding to a distance distribution $\varphi v(r)$. Therefore, the intensity without the collimation effect $[\Psi_{\upsilon}(q)]$ can be expressed as:

$$\Psi_{\upsilon}(q) = T_1 \varphi_{\upsilon}(r) \tag{4.8}$$

The smeared intensity (i.e., after adding the collimation effect) can be obtained after execution of T_2, T_3, and T_4 to give:

$$\chi_{\upsilon}(q) = T_4 T_3 T_2 \Psi_{\upsilon}(q) = T_4 T_3 T_2 T_1 \varphi_{\upsilon}(r) \tag{4.9}$$

Hence, $\chi(q)$ represents the approximated scattering data from a particle with D_{max}. The next step is the stabilization of these coefficients. The stabilized least squares conditions are given below.

$$L + \lambda_L N_{c'} = \text{minimum} \tag{4.10}$$

$$L = \frac{\int_{q_1}^{q_2} \left[I_{\exp}(q) - \sum_{\upsilon=1}^{N} c_{\upsilon} \chi_{\upsilon}(q) \right]^2}{\sigma^2(q)} dq \tag{4.11}$$

$$N_{c'} = \sum_{\upsilon=1}^{N-1} (c_{\upsilon+1} - c_{\upsilon})^2 \tag{4.12}$$

Here, q_1 and q_2 are the first and last data points, respectively, $I_{\exp}(q)$ is the experimental intensity, σ^2 is the estimated variance of the observed intensity, and λ_L is the stabilization parameter or Lagrange multiplier [5–7]. The optimum fit of the observed data points is given by:

$$I_A(q) = \sum_{\upsilon=1}^{N} c_{\upsilon} \chi_{\upsilon}(q) \tag{4.13}$$

$I_A(q)$ represents the approximated scattering curve, which should be similar to $I_{\exp}(q)$. Therefore, it can be concluded that the approximated distance distribution function $[p_A(r)]$ represents the $p(r)$ of the experimental curve.

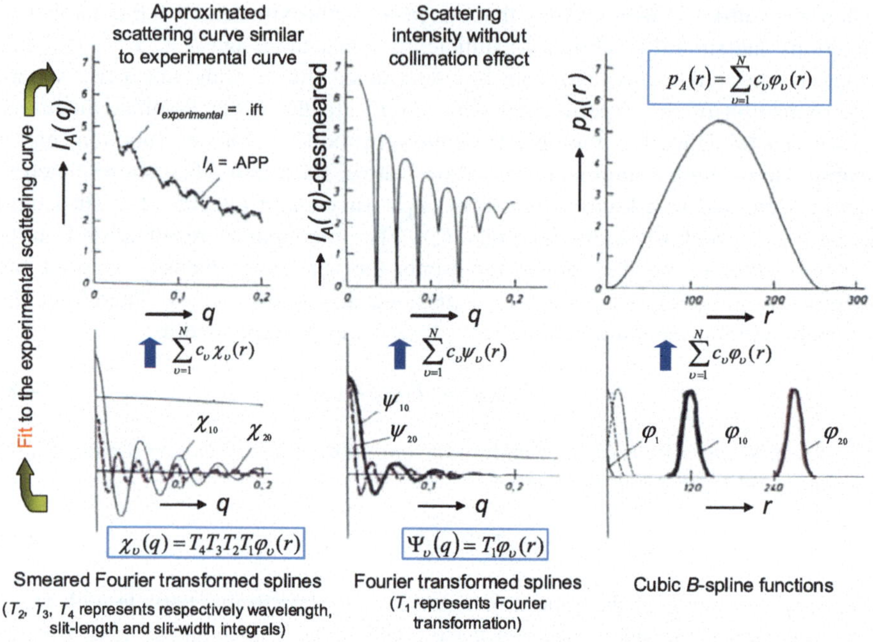

Fig. 4.2 The representative diagram of generalized indirect Fourier transformation (GIFT) method for spherical particle. Partially reproduced from [5] with permission

Figure 4.2 shows a representative diagram of the GIFT method for spherical particles, where $p(r)$ consists of N cubic B-spline functions and directly shows the probability of finding a pair of electron densities at a particular distance r [5]. Therefore, each spline function can be considered as the distance between a pair of dispersed NPs, for example stacked silicate layers. An advantage of the GIFT method is that the form and structure factors can be determined simultaneously from the measured scattering data with a correction for the instrumental broadening effect [8–10]. Therefore, to evaluate $p(r)$ from the scattering curve, one has to consider the values of N, D_{max}, and λ_L. If there is a difference between $I_A(q)$ and $I_{exp}(q)$, then it is necessary to consider the effect of the structure factor.

D_{max} does need not to be a precise estimation of D. As a rough estimate, $D_{max} \leq (\pi/q_1)$, where q_1 is the lowest scattering angle. A theoretical limitation for the number of functions $N = N_{max}$ follows from the sampling theorem. A main concept of the indirect Fourier transformation technique is to start with a large number of coefficients to guarantee a sufficient representation of the distance distribution function. To approximate, $N_{max} \leq (q_2 \cdot D_{max}/\pi)$, where q_2 is the maximum scattering angle. The stabilization parameter restricts the oscillation of the spline functions (i.e., oscillation of $p(r)$) and should be chosen so that the approximated scattering curve (determined on the basis of the $p(r)$) is similar to the experimental scattering curve. Only then can it be concluded that the $p(r)$ related to the approximated scattering curve is the same as that for the experimental scattering curve.

The structure factor is determined using the GIFT method with the modified Caillé theory for lamellar phases. There are two theories applicable to lamellar systems. Firstly, the paracrystalline theory, a general theory for disorder of the first and second kind was developed by Hosemann and Bagchi [11] and Guinier [12]. This was the first attempt to address the disorder in multilamellar arrays. The paracrystalline theory of the first kind assumes that there are stochastic distance fluctuations around the well-defined mean layer positions of equal separation; i.e., the long-range order is maintained. The paracrystalline theory of the second kind describes fluctuations of bilayer separations relative to the nearest neighbors of ideally flat bilayers. These fluctuations are not correlated and the long-range periodic order collapses [12]. However, the Caillé theory developed on the basis of the thermodynamic theory of DeGennes for smectic liquid crystals is preferable as it considers the bending of bilayers in addition to fluctuations in the mean spacing between them [13]. The modified Caillé theory proposed by Zhang et al. considers the finite size of the lamellar stack [14, 15]. This modification does not affect the quantitative results obtained by the original Caillé theory, but the modification is necessary to obtain better quantitative fits to the data and particularly for extracting the correct form factor, which could be used later to obtain an electron density profile.

As long as the bilayer is unilamellar, there exists a direct relationship between the electron density profile in the perpendicular direction to the midplane of the bilayer and the form factor. The lateral arrangement of multilamellar bilayers is represented by the structure factor and can be determined by either the paracrystalline or Caillé theory with a few parameters. It is necessary to assume either a form factor or a structure factor to evaluate the scattering data using the paracrystalline or modified Caillé theory. Frühwirth et al. [16] implemented the modified Caillé theory with GIFT to analyze stacked lamellar systems. This model is defined by three parameters: the number of coherently scattering bilayers (n); the repeat distance (d) of bilayer; and the Caillé parameter (η_1). According to the modified Caillé theory, the structure factor can be expressed as:

$$S(q) = n + \left\{ 2\sum_{m=1}^{n-1} (n-m)\cos(mqd)\exp\left[-\left(\frac{d}{2\pi}\right)^2 q^2 \eta\gamma\right] (\pi m)^{-(d/2\pi)^2 q^2 \eta_1} \right\} \tag{4.14}$$

where γ is Euler's constant (=0.5772). The parameter η_l can be expressed as:

$$\eta_1 = \frac{q_1^2 k_B T}{8\pi (K_c B)^{1/2}} \tag{4.15}$$

where,

$$q = (2\pi/d) \tag{4.16}$$

K_c is the bending modulus and B is the bulk modulus for compression. Since the two moduli cannot be determined independently from the scattering data, one can consider η_1 as a measure of flexibility of the bilayers. According to the author, increasing the number of bilayers results in higher intensity and narrower peaks, and increasing the Caillé parameter leads to a faster decay of the peaks of higher order.

The electron density for the lamellae can be written as follows (assuming that the lamellae are homogeneous along the basal plane).

$$\rho(r) = \rho_0 \cdot \rho_t(x) \tag{4.17}$$

Here, ρ_0 is a constant and x is the normal distance from the central plane in the lamellae. Therefore, $\rho_t(x)$ represents the electron density along the thickness cross-section profile [5]. There are two ways to determine the electron density profile. In the conventional method, the scattering amplitude is determined from the scattering intensity by a simple square root operation. However, the main challenge is determining to correct sign (the so-called phase problem). The second method is estimating the electron density from the distance distribution function by a convolution square root technique. This method does not suffer from the phase problem. Hosemann and Bagchi [11] and Engel [12] showed that for the lamellar system, the convolution square root has a unique solution (except for a factor ± 1) if the function has a finite range of definition and the function is symmetrical [5]. Glatter used the convolution square root method in a different way. He deconvoluted the approximated electron density distribution in order to obtain the distance distribution function for highly symmetric systems (sphere, cylinder, or lamella). The electron density was approximated in its range of definition by a linear combination of a finite number of functions that have to be linearly independent in this range, expressed as follows:

$$\bar{\rho}(r) = \sum_{i=1}^{N} c_i \varphi_i(r) \tag{4.18}$$

where N is the number of functions, r is the normal distance from the center of symmetry, $\varphi_i(r)$ is the equidistant step function (cubic B-spline of zero order) with a width ΔR allowing analytical integration of the overlap integrals, and c_i is the height of the step functions. Equation (4.18) corresponds to a nonlinear distance distribution function (see (4.19)), which can be solved in an interactive stabilized manner to describe the $p(r)$ function obtained from IFT/GIFT methods [17–19].

$$\bar{p}(r) = \sum_{i=1}^{N} V_{ii}(r)c_i^2 + \sum_{i>k} V_{ik}(r)c_ic_k \tag{4.19}$$

If there is some deviation from high symmetry, which is known as the polydispersity of a sample, then the $p(r)$ determined by the deconvolution (DECON) method (performed using DECON software) will be slightly different than the value determined from GIFT. By estimating the amount of polydispersity, a good match between the $p(r)$ values determined by GIFT and DECON can be achieved. Therefore, the electron density distribution derived from DECON should represent the experimental scattering curve.

Although XRD is used widely to analyze the dispersion of nanoclays in polymer nanocomposites, SAXS is beneficial for probing the dispersion characteristics of highly delaminated structures [1, 20–23]. Additionally, as discussed above, GIFT allows detailed investigation on the inter-particle correlation function and internal structure from the electron density distribution [24, 25]. SAXS analysis has demonstrated that the filler loading plays a vital role in controlling the network structure of dispersed silicate layers in a polymer matrix [24]. The current authors have extensively exploited this method for clay-containing poly[(butylene succinate)-co adipate] (PBSA) nanocomposites and determined the percolation threshold concentration of the nanoclay [24]. The $p(r)$ values of various nanocomposites obtained from GIFT analysis are shown in Fig. 4.3 [24]. The regions with opposite signs of different electron density give negative contributions to $p(r)$. The r-value at

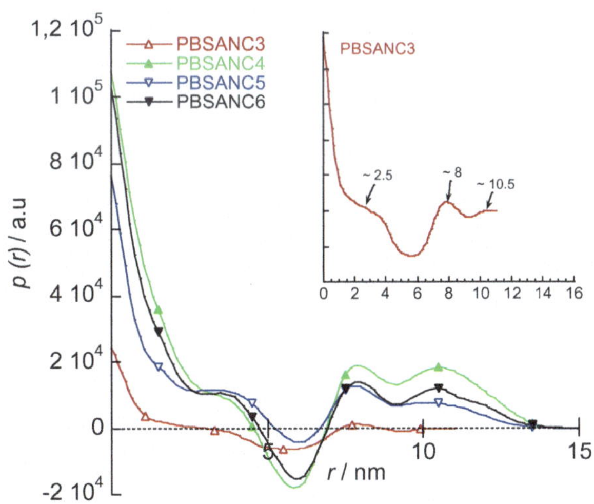

Fig. 4.3 The pair-distance-distribution function, $p(r)$ for nanocomposites showing the probability of finding neighboring particles in systems with increase in clay concentration. PBSA nanocomposites (PBSANCs) with four different C30B loadings of 3, 4, 5, and 6 wt%, were abbreviated as PBSANC3, PBSANC4, PBSANC5, and PBSANC6, respectively. Reproduced with permission from [24]. Copyright 2010, Elsevier Science Ltd

which $p(r)$ drops to zero indicates the largest single particle dimension. As evidenced from the figure, the number of correlation maxima (peaks) increases with an increase in nanoclay loading. These correlation maximums represent the average radial distance to the next neighboring domain, commonly known as long spacing. When the neighbors overlap, the peaks do not possess a tail; rather, the curve shows a maxima and minima.

The electron density profiles obtained from the $p(r)$ values (see Fig. 4.4) showed that the nanocomposites had a core–shell particle structure [24]. When the clay platelets started delaminating in the nanocomposites, these core-shell structures start to grow with the previously peeled nanoclay layers as a shell. PBSANC3 had a core thickness of approximately 6 nm and a shell layer thickness of 2.6 nm. In the case of PBSANC4, the core thickness decreased and the total shell thickness for the two shells increased dramatically. There was no remarkable subsequent change in core thickness or total shell thickness for either PBSANC5 or PBSANC6. Therefore, the percolation threshold value of C30B loading in the case of C30B-containing PBSANCs was 5 wt%.

The SAXS spectra and electron density profile obtained from the *in situ* temperature-assisted SAXS can elucidate the molecular dynamics mechanisms of nanocomposite formation [26]. It was shown that the absorbed moisture of fumed silica had a plasticizing effect on the polycarbonate (PC) matrix. Plasticization advanced the segmental movement of PC chains in the PC/fumed silica composite above 150 °C.

The number of aggregated particles (N_{agg}) dispersed in a polymer matrix can be determined according to (4.20) [27]:

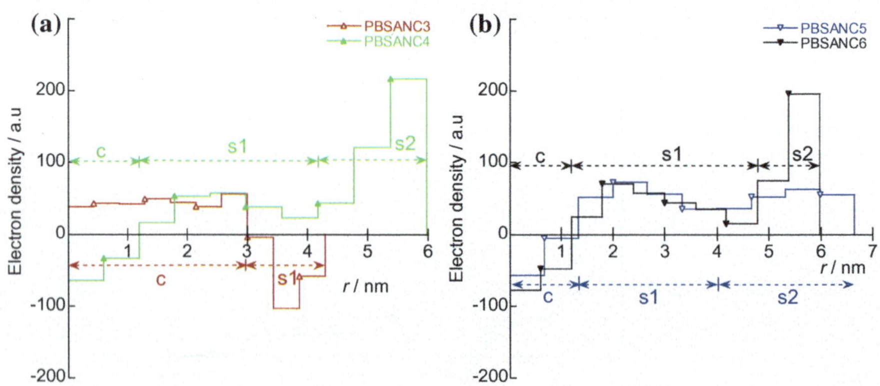

Fig. 4.4 Electron density profile of the model nanocomposites: **a** for PBSANC3 and PBSANC4 and **b** for PBSANC5 and PBSANC6. PBSA nanocomposites (PBSANCs) with four different C30B loadings of 3, 4, 5, and 6 wt%, were correspondingly abbreviated as PBSANC3, PBSANC4, PBSANC5, and PBSANC6, respectively. Here the core is denoted by 'c', inner shell by 's1' and outer shell by 's2'. Reproduced with permission from [24]. Copyright 2010, Elsevier Science Ltd

$$N_{\text{agg}} = \left(\frac{2\pi}{q^*}\right)^3 \left(\frac{\varphi}{V}\right) \tag{4.20}$$

where q^*, φ, and V are the scattering peak position, volume fraction, and volume of the particle, respectively.

4.2.1.2 Orientation of Dispersed Nanoparticles in Polymer Nanocomposites

X-ray scattering has been extensively used to estimate the gallery spacing of nanoclay platelets dispersed in polymer nanocomposites. The disappearance of a discrete peak is attributed to exfoliation. However, such interpretation might be misleading sometimes. The preferred in-plane orientation of platelet-type nanoclays in the nanocomposite can drastically reduce the scattering intensity, as shown in the Fig. 4.5 [23, 28]. The schematic diagrams (Fig. 4.5a, b) describe the interaction of

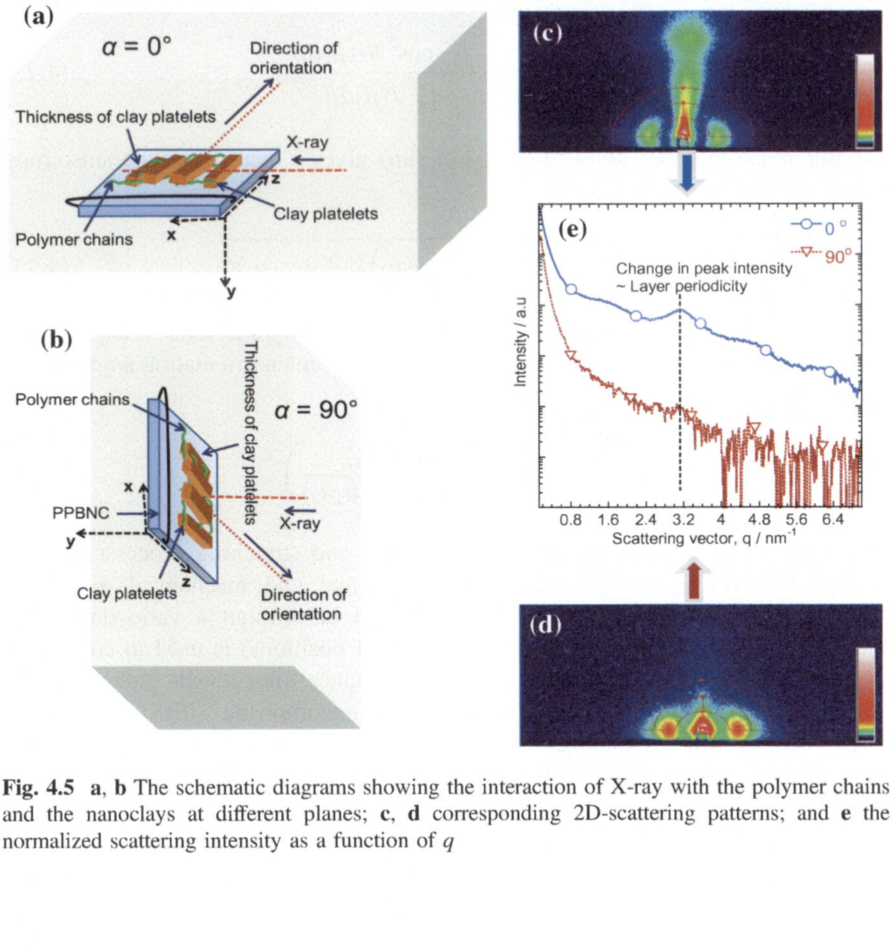

Fig. 4.5 **a**, **b** The schematic diagrams showing the interaction of X-ray with the polymer chains and the nanoclays at different planes; **c**, **d** corresponding 2D-scattering patterns; and **e** the normalized scattering intensity as a function of q

X-rays with polymer chains and nanoclays in different planes. The 2D-scattering patterns (Fig. 4.5c, d) and the normalized scattering intensity as a function of q (Fig. 4.5e) show that the intensity and appearance/disappearance of a peak depends on the angle of interaction between the X-ray and particle.

In a densely packed system of particles, the positional ordering can develop a preferential orientation, especially when particles are not spherical. The degree of orientation can be easily detected from 2D SAXS patterns. Usually, an arc-profile is used to determine the orientation of crystals in a certain basal plane. A point on the azimuthal scan can be presented by a unit vector, u, such that $u_1 = \cos\beta$ and $u_2 = \sin\beta$; where β is the azimuthal angle. The anisotropy in the X-ray scattering pattern can be obtained from the weighted average of the second moment tensor of u following:

$$\langle uu\rangle = \begin{bmatrix} \langle u_1u_1\rangle\langle u_1u_2\rangle \\ \langle u_1u_2\rangle\langle u_2u_2\rangle \end{bmatrix} = \begin{bmatrix} \langle\cos^2\beta\rangle\langle\sin\beta\cos\beta\rangle \\ \langle\sin\beta\cos\beta\rangle\langle\sin^2\beta\rangle \end{bmatrix} \tag{4.21}$$

Here, $\langle\cdots\rangle$ represents an average weighted by the azimuthal intensity distribution and e.g., $\langle\cos^2\beta\rangle$ can be expressed as,

$$\langle\cos^2\beta\rangle = \frac{\int_0^{2\pi}\cos^2\beta I(\beta)d\beta}{\int_0^{2\pi}I(\beta)d\beta} \tag{4.22}$$

The difference in eigenvalues $(\lambda_1 - \lambda_2)$ of $\langle uu\rangle$ gives a measure of the anisotropy factor and can be expressed as:

$$\lambda_1 - \lambda_2 = \sqrt{(\langle u_1u_1 - u_2u_2\rangle)^2 + 4\langle u_1u_2\rangle^2} \tag{4.23}$$

The degree of anisotropy determined by the software program 'tdoa' is $(\lambda_1 - \lambda_2)$ in percent. The mean orientation angle or the average domain orientation angle can be expressed as:

$$\bar{\chi} = \frac{1}{2}\tan^{-1}\left(\frac{2\langle u_1u_2\rangle}{\langle u_1u_1\rangle - \langle u_2u_2\rangle}\right) \tag{4.24}$$

The sample can be oriented either due to shear and stretching processes during polymer processing or upon experiencing thermal and mechanical treatments [28–30]. Conventionally, point collimation with or without a vario-stage (that enables tilting, rotation, and scanning at different positions) is used to collect the scattering spectrum for such analysis. At an instance after tensile stretching, the orientations of pure polymer (PBSA) and nanocomposite (PBSANC3) were determined by tilting, rotating, and scanning the specimens with respect to the incident X-ray beam [28]. It was shown that the dispersed clay platelets were oriented along the z-axis, i.e., the direction of elongation during tensile testing. The second observation was that the thicknesses of the clay layers were in the xy- and yz-planes. The surface of the clay platelets were in the xz-plane (see Fig. 4.5a, b). In addition to determining particle dispersion and distribution, SAXS instruments fitted with Linkam shearing or stretching devices can analyze the time dependent structural changes during stress relaxation and hysteresis [31]. The authors observed that the nanoclay orients almost instantaneously, while the polymer chains become elongated in the stretching direction, followed by delamination off the polymer chains that were adsorbed on the nanoclays.

Processing conditions, such as variations in the feeding rate during extrusion, can affect the orientation of the nanoclay platelets in the polymer nanocomposite [32]. Figure 4.6 shows that increasing the feed rate resulted in a random distribution of nanoclay in the nanocomposite, while a slower feed rate allowed orientation of the nanoclay platelets to occur. The shearing device coupled with the SAXS system elucidated changes in the orientation angle and anisotropy factor as a function of shear rate [33]. In addition, compression can also introduce anisotropy, e.g., in the blend composite of polypropylene (PP)/ethylene-propylene-diene terpolymer rubber (EPDM)/Cloisite® 15A (C15A) [34].

4.2.1.3 Lamellar Crystal Thickness

SAXS can also be employed to estimate the lamellar crystal thickness from the first-order long period. To achieve this, the first- and second-order reflections are

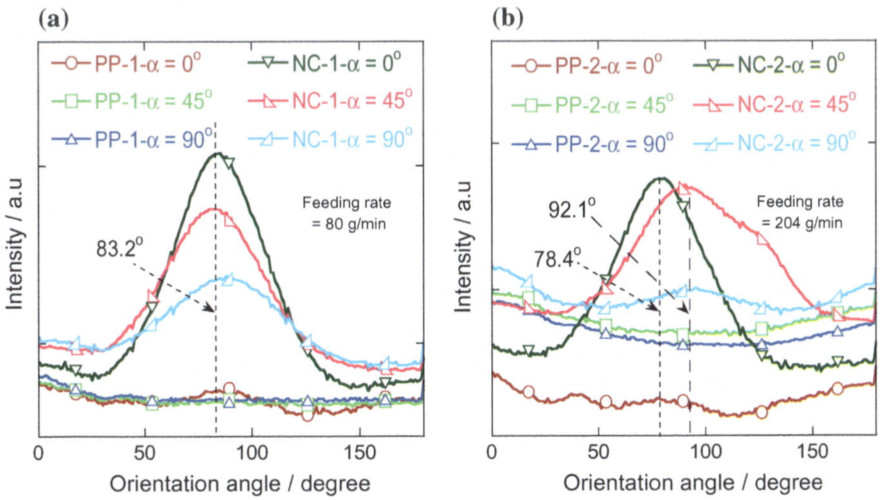

Fig. 4.6 Change in orientation of clay platelets with the feeding rate **a** NC-1 and **b** NC-2. Feed rate used for NC1 and NC2 were 80 and 204 g/min, respectively

separated by fitting the scattering profile with the product of Gaussian and Lorentzian functions [35]. Then, the first- (l_1) and second (l_2)-order long periods can be estimated using Bragg's law. The lamellar crystal thickness (l_c) can then be estimated as follows:

$$l_c = \varphi l_1 \tag{4.25}$$

where, φ is the crystallinity of the material at a particular condition. As an estimation, the chain dimension in the melt is the radius of gyration (R_g), the condition for crystallization without disentanglement is:

$$l_c \leq R_g \tag{4.26}$$

Fu et al. [36] assumed that if the crystal is smaller than the chain dimension (i.e., satisfying (4.26)), disentanglement may not be required for crystallization from the melt state. The entanglements can be shifted toward the amorphous region, which eventually can form stereo-defects as noncrystallizable entities. However, if $l_c > R_g$, crystallization occurs via disentanglement of polymer chains. For Gaussian chains, R_g can be estimated from (4.27) [35]:

$$R_g^2 = \frac{R_0^2}{6} \tag{4.27}$$

where, R_0 is the mean squared end-to-end distance and can be calculated using the characteristic ratio C_∞, sum of the square of the length of backbone bonds in one monomer unit (a_b^2), and degree of polymerization (N), as shown below [35]:

$$R_0^2 = C_\infty a_b^2 N \tag{4.28}$$

The long period (and hence, lamellar thickness) as a function of nanoclay concentration in a polyamide 6(PA6)/NanomerI.30TC composite showed that the crystalline morphology of PA6 changed dramatically in the presence of the nanoclay [20].

4.2.1.4 Particle Dimension and Specific Surface Area

R_g is determined from the Guinier equation below:

$$I(q) = (\Delta\rho)^2 V^2 \exp\left(-\frac{q^2 R_g^2}{3}\right) \tag{4.29}$$

R_g does not contain any information about the shape and internal structure of the particle [4]. However, by knowing R_g and the form factor of scattering [$P(q)$], it is possible to estimate the average radius of the particle. The value of [$P(q)$] can be approximated by a Gaussian curve at small angles, where the curvature of the Gaussian depends on the overall size of the particle [4]. In the case of spherical particles,

$$P(q) \approx a_0 \exp\left(\frac{-R_g^2}{3} q^2\right) \tag{4.30}$$

The parameter a_0 can be extrapolated to zero angle intensity and a_0 = 1 to the zero angle position. Since the intensity is directly proportional to $P(q)$,

$$\ln[\Delta I(q)] = \ln[a_0] - \frac{R_g^2}{3} q^2 \tag{4.31}$$

Therefore, R_g and a_0 can be determined from $\ln[\Delta I(q_0)]$ versus q^2 curves (a Guinier plot). Similarly, the radius of gyration of the cross-section of cylindrical particles (R_C) and the thickness radius of gyration of lamellar particles (R_T) can be determined from (4.31) and (4.32), respectively [4].

$$qP(q) \approx a_0 \exp\left(\frac{-R_C^2}{2} q^2\right) \tag{4.32}$$

$$q^2 P(q) \approx a_0 \exp\left(-R_T^2 q^2\right) \tag{4.33}$$

Multiplying $P(q)$ by q (see 4.31) eliminates the contribution of the axial dimension (assumed to be infinitely long). Similar logic can be applied for the lamellar structures; hence, $P(q)$ is multiplied by q^2 in 4.32. Initial slopes of 0, −1, and −2 indicate globular, cylindrical, and lamellar-shaped particles, respectively (see the log [$P(q)$] vs. log q plot shown in Fig. 4.7). The average radius for spherical and cylindrical particles and the half thickness for a homogeneous plate can be determined from (4.34) to (4.36).

$$\textit{Radius of spherical particle} = \sqrt{\frac{5}{3}} R_g \tag{4.34}$$

$$\textit{Radius of cylindrical particle} = \sqrt{2} R_C \tag{4.35}$$

$$\textit{Half thickness of platelets} = \sqrt{3} R_T \tag{4.36}$$

The specific surface area and interfacial area can also be estimated using SAXS [37]. The intensity of the tail region of the scattering pattern is approximated by Porod's law:

$$\lim_{q\to\infty}\left[I(q)q^3\right] = \frac{(\Delta\rho)^2 S}{16\pi^2} \tag{4.37}$$

Here, S is the surface area of the particle and can be determined from (4.37) by knowing the absolute intensity. Equation (4.37) is independent of the dimension, shape, and porosity of the particle and valid for a concentrated system where interfaces between the particles have little effect on the tail of the SAXS spectrum. However, measurement of the absolute intensity in order to determine S can be avoided by considering the invariant Q:

$$Q = \int_0^\infty 2\pi q I(q)dq = (\Delta\rho)^2 V \tag{4.38}$$

Therefore, (4.37) reduces to

$$\lim_{q\to\infty}\left[I(q)q^3\right] \cdot \frac{16\pi^2}{Q} = \frac{S}{V} \tag{4.39}$$

The correlation between S and R_g for nanoclay-containing composites of natural rubber (NR) and styrene butadiene rubber (SBR) composites is presented in Fig. 4.8 [37]. As evidenced from the figure, the interfacial area decreases with increasing size of the agglomerated/stacked nanoclays.

SAXS can also be employed to determine the surface per mass (S_M) according to (4.40) and (4.41).

$$S_M = 1000\pi \frac{\varphi(1-\varphi)}{d}\frac{K}{Q} \quad \text{(for desmeared scattering data)} \tag{4.40}$$

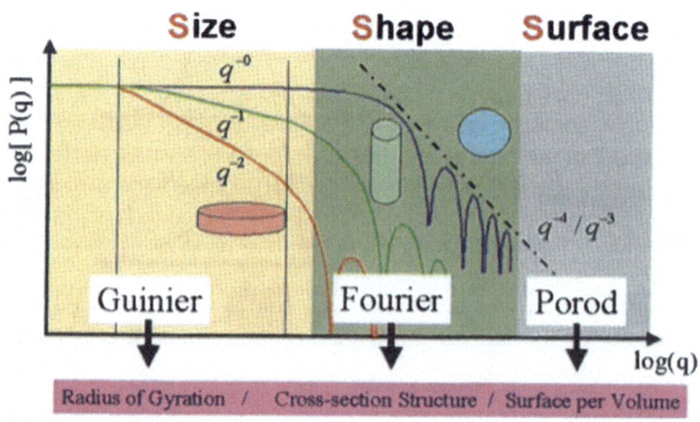

Fig. 4.7 The information domain of a particle form factor. Reproduced with permission from [4]. Copyright 2006, Anton Par GmbH

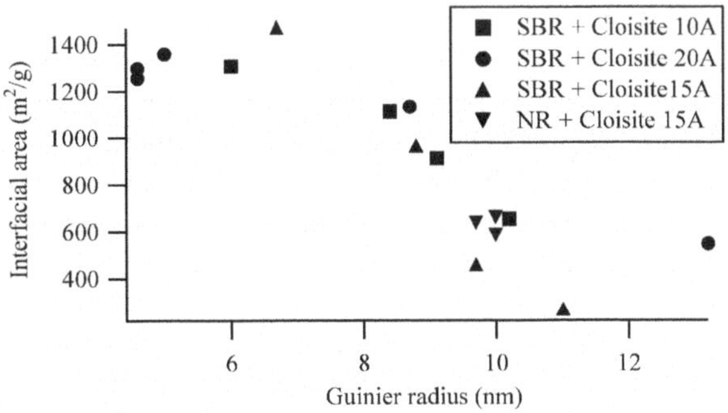

Fig. 4.8 Trend of the matrix/clay interfacial area as a function of the radius of gyration of clay particles. Reproduced with permission from [37]. Copyright 2012, the American Chemical Society

$$S_M = 4000\pi \frac{\varphi(1-\varphi)}{d}\frac{K}{Q} \quad \text{(for smeared scattering data)} \tag{4.41}$$

Here, φ, d, and K represent the volume fraction, density, and intercept of the Porod plot, respectively. The invariant, Q can be expressed as the sum of Q_1 (Guinier extrapolation), Q_2 (experimental), and Q_3 (Porod extrapolation). Furthermore, correlation between the interfacial surface area and the moduli (see Fig. 4.9) shows that the stiffness/tensile modulus of the composite increased with increasing total interfacial area [37].

4.2.2 *X-ray Diffraction*

In principle, X-ray diffraction (XRD) is fundamentally similar to the SAXS; both techniques are used to extract structural information. However, SAXS provides

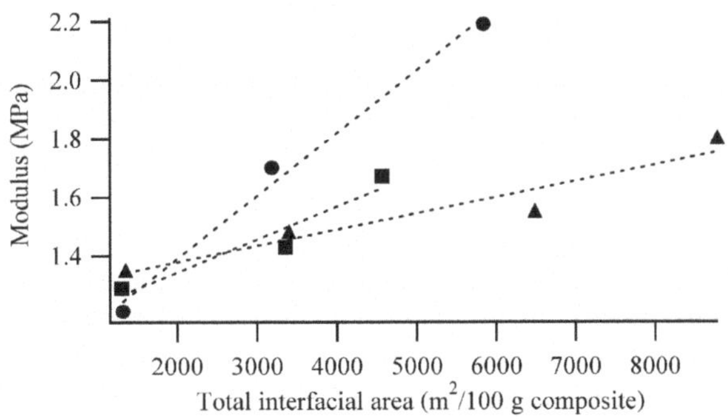

Fig. 4.9 Elastic modulus as a function of interfacial area, circles represents the NR-based composites, square represent SBR10 composite series and triangles indicate SBR20 composite series. The lines were drawn to show the trend. Reproduced with permission from [37]. Copyright 2012, the American Chemical Society

information related to the particle as a whole, while XRD probes atomic arrangements (crystal structures) inside the particle. Therefore, combining both SAXS and XRD data is very powerful for analyzing multi-component systems such as polymer composites. XRD is mainly used to determine the atomic structure of a crystal. The incident X-ray interacts with the atoms and is diffracted in specific directions. Understanding the crystal structure is important for e.g., identifying compounds, ores, and polymorphs, and analyzing material fatigue; it is important to note that the integral intensity depends on the volume fraction of the phases, crystal structure (structure factor), lattice vibration (temperature factor), composition and ordering in the solid solution, aggregate structure (microabsorption), orientation distribution (texture factor), and lattice defects [38]. Similarly, the peak position depends on the composition in the solid solution, thermal expansion, and internal stresses (kind I) [38]. The shape of the peak (broadening) depends on the internal stresses (kind II and III), particle size, stacking faults, and dislocation substructure [38].

In PNC research, XRD is an important tool for probing the dispersion of the NPs in the polymer matrix. XRD is also used to determine the crystal lattice and size of the polymer crystal. Properties of the polymer, such as the transparency, and the mechanical, thermal, and barrier properties can be directly correlated with the shape and size of the crystals. Smaller crystallite sizes generally have better transparency and mechanical performance, while larger spherulites (crystals) sometimes can impede the permeation of gas molecules [39].

4.2.2.1 Dispersion Characteristics

XRD has been employed to investigate the dispersion of different types of NPs (e.g., nanoclays, graphene, and metal oxide particles) dispersed in various polymer matrices [40–44]. The *d*-spacing between the two crystal planes is determined by Bragg's law:

$$n\lambda = 2d \sin\theta \tag{4.42}$$

where, n, λ, d, and θ represent a positive integer, wavelength of the incident X-ray, *d*-spacing of the crystal, and the scattering angle, respectively.

Figure 4.10 shows XRD patterns of nanoclay and PP nanocomposites in the small-angle region [32]. In both nanocomposites (NC-1, PNC processed at a feeding rate of 80 g/min and NC-2, PNC processed at a feeding rate of 204 g/min), the ratio of PP:PP-g-MA(maleic anhydride grafted PP):silicate content of clay was kept at 97:9:3. Two different feeding rates (80 and 204 g/min) were used during the dilution of the PP-g-MA/Cloisite® C20A (C20A) master batch in PP. The figure shows a characteristic peak of the C20A nanoclay at 3.6°, corresponding to a $d_{(001)}$-spacing of 2.45 nm. In NC1 and NC2, this peak shifted towards smaller angles, appearing at 2.5° and 2.7°, respectively (corresponding to $d_{(001)}$-spacings of 3.61 nm and 3.29 nm, respectively). Such an increase in the *d*-spacing indicates that the polymer chains were intercalated in the clay gallery. Another interesting

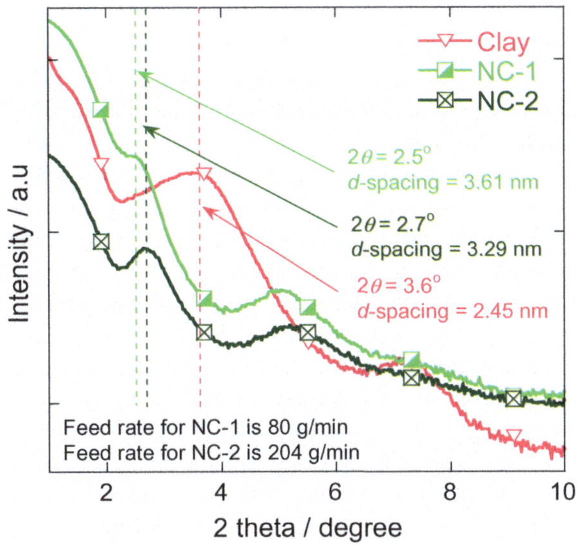

Fig. 4.10 X-ray diffraction patterns of compression molded polypropylene nanocomposites in the small angle region

observation is that the characteristic XRD peak was weaker for NC1 than for NC2, even though they contain almost the same amount of silicate. Hence, parallel stacking of the dispersed clay was lower in NC1 than in NC2.

In addition to the dispersion of nanoclays, XRD can provide information about the interaction of the nanofiller with the polymer matrix. Shanthala et al. [45] observed an amorphous halo from polypyrrole (PPy). However, many small crystal peaks appeared for a PPy/copper zinc iron oxide nanocomposite due to interaction of the NPs with the polymer via chemical bonding or interaction with macromolecular chains of the matrix. Furthermore, Zhu et al. [41] showed that when graphene oxide (GO) nanosheets absorb carrageenan (Car) macromolecules through hydrogen-bond interactions, the interlayer spacings between GO nanosheets increased.

4.2.2.2 Analysis of Crystal Phase

Different crystal structures of the same material are important for many applications. For example, the presence of both anatase and rutile phases of TiO_2 in polyaniline (PANI) nanocomposites plays an important role in improving the photocatalytic activity [46]. The synergistic effect of both anatase and rutile phases inhibit the recombination of the electron-hole pair. Identification of crystal structure is also important to predict the mechanical properties. For example, in the presence of β-crystals, isotactic PP exhibits improved elongation at break, and higher toughness and impact strength, while the α-polymorph shows increased stiffness

[47, 48]. Therefore, a combination of α- and β-crystals can provide optimum toughness and stiffness.

In the case where all samples have the same surface roughness and thickness, the optical transparency depends on the crystallinity. When the overall crystallinity is similar, the amount of -phase could be the controlling factor. De Santis and Pantani [49] demonstrated the dependence of the crystallinity on the cooling rate and the number of recycling steps. It is interesting to note that while the α-phase crystallinity decreased with increasing cooling rate, the mesomorphic phase fraction increased with the number of recycling steps. Moreover, the authors correlated the birefringence and opacity of the PP samples before and after recycling steps with the amount of α-phase present in the samples. Birefringence is an optical phenomenon where polymer exhibits different refractive indices for light with plane polarization in two perpendicular directions (ordinary and extraordinary rays). According to the reported results, the birefringence and hence, opacity of the polymer, increased with increasing α-phase crystallinity. Such results are important for product development. The phase and its degree of crystallinity also play important roles in the gas barrier performance of polymers [39]. The degree of crystallinity (X_c) can be determined from the intensity of the crystalline peaks (I_c) and the amorphous region (I_a).

$$X_c = [I_c/(I_c + I_a)]\,.\,100 \tag{4.43}$$

Malas et al. [50] established a correlation between the increase in the glass transition temperature and degree of crystallinity, and the dielectric strength of the interfacial polarization for a polyethylene oxide and reduced graphene oxide system. The authors found that the strength of the interfacial entrapment of charges during dielectric spectroscopy is related to reduced mobility of the polymer chains.

Another important crystal parameter influencing the properties of polymeric composites is the crystallite size, L, which can be estimated by the well-known Scherrer equation:

$$L = \frac{K\lambda}{\beta \cos\theta} \tag{4.44}$$

Here, K, λ, β, and θ are the Scherrer constant ($\approx$1), wavelength of the incident X-ray beam, full-width-at-half-maximum of the peak, and the scattering angle, respectively. Subsequently, the number of crystallites per unit area (C) in the film can be estimated as follows [51].

$$C = \frac{t/L^3}{unit\ area} \tag{4.45}$$

Various X-ray diffraction analysis methods (such as pole figure construction and Fourier transform peak shape deconvolution) can be employed to quantify texture changes, and the relative degree of crystallinity and lattice order. The texture coefficient (TC) of each (*hkl*) plane is determined from the XRD spectrum using (4.46) [52].

$$\mathrm{TC}_{(hkl)} = \frac{I_{(hkl)}}{I_{0(hkl)}} \left[\frac{1}{N} \sum_{i=1}^{N} \frac{I_{(h_i k_i l_i)}}{I_{0(h_i k_i l_i)}} \right]^{-1} \tag{4.46}$$

where $I\ (h_ik_il_i)$ is the intensity of the $(h_ik_il_i)$ diffraction peak of the sample under investigation; $I_0\ (h_ik_il_i)$ is the intensity of the $(h_ik_il_i)$ plane of a completely random sample taken from a powder diffraction file (PDF) card; and N is the number of diffractions considered in the analysis. For randomly oriented crystals, $\mathrm{TC}_{(hkl)} = 1$ [53]. Higher values indicate more grains oriented in a given (*hkl*) direction.

4.3 Analysis in Real Space

4.3.1 Electron Microscopy

In electron microscopy (EM), an electron beam is used for imaging, where the interaction between the electron beam and the sample produces various scattering signals. Different electron microscopy techniques, such as transmission EM (TEM), scanning TEM (STEM), and focused ion beam scanning EM (FIB-SEM) allow qualitative analysis of the internal structure via direct visualization. Some microscopy techniques require sample preparation involving either a microtome or milling by ion-beams. In the case of SEM, two types of signals are usually detected; secondary electrons and backscattered electrons. The secondary electrons originate from the surface, or close to surface, and are due to inelastic interaction between the primary electron beam and atoms in the samples. These are beneficial for imaging the surface topography. In contrast, backscattered electrons originate due to elastic collisions of electron beam with the atoms. As the probability of elastic collision increases with increasing size of the atom, larger (heavier) atoms produce stronger scattering signals than smaller (lighter) atoms. Hence, the backscatter signal is directly proportional to the atomic number. Therefore, backscattered signals are useful for detecting different phases or compositions in the sample.

The dispersion characteristics of nanoclays in PBSANC3 after tensile stretching are shown in Fig. 4.11 [54]. Further image analysis by energy dispersive spectroscopy (EDS) confirmed that the dark features were nanoclay platelets which were well-oriented in the direction of the applied tensile strain. To gain a deeper understanding of the degree of dispersion of the nanoclays in the polymer matrix, EM images can be further analyzed by image processing. STEM images and the image analysis process are depicted in Fig. 4.12, showing that the nanoclay loading plays a vital role in controlling the network structure [24]. These results also showed that 5 wt% nanoclay was the percolation threshold for forming a strongly flocculated structure of dispersed silicate layers. The authors also reported that the

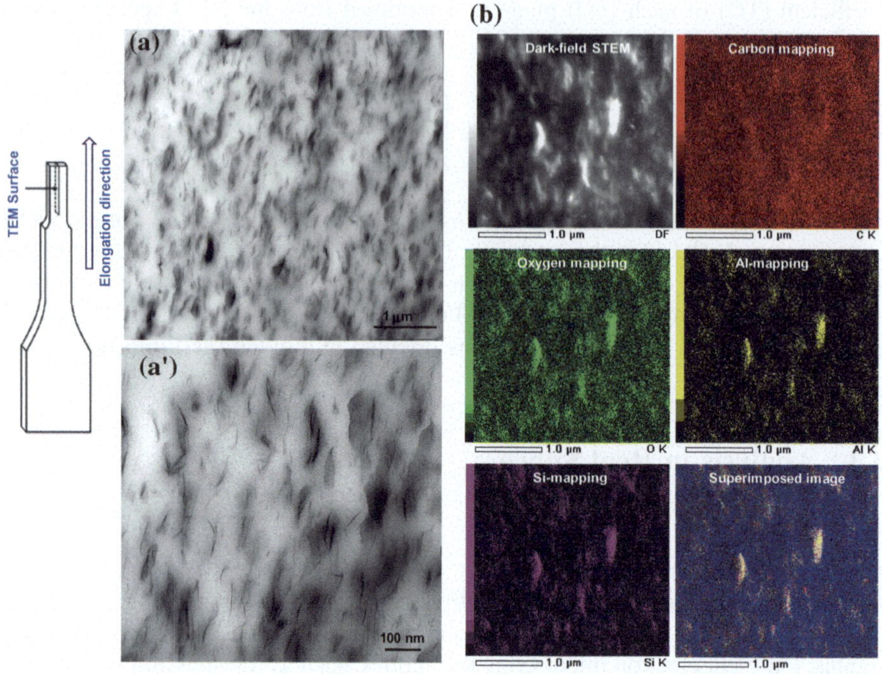

Fig. 4.11 **a**, **a′** Bright-field TEM image of nanocomposite sample after tensile test and **b** X-ray mapping of various elements in scanning transmission electron microscopy mode. This result shows silicate layers are oriented towards the direction of applied strain. Reproduced with permission from [54]. Copyright 2010, Elsevier Science Ltd

dispersion characteristics of nanoclays influence the melt-flow behaviors of nanocomposites.

SEM is commonly used to investigate the properties of e.g., fracture surfaces and the morphology of polymer blends. SEM equipped with EDS provides elemental information of the nanoparticle dispersed in the PNCs. Depending on the contrast of the electron density; SEM can also be used to determine the thickness of individual layers present in the multilayered structure.

3D-tomography using a FIB-SEM cross-beam system can be used to analyze the orientation of NPs (e.g., nanoclays) dispersed in a polymer matrix. Such analysis is complementary to SAXS analysis. The process involves cutting 2D-slices through a selected volume by ion-beam milling, then imaging the cross-section using high-resolution SEM (HR-SEM). The 2D-images are then aligned by cross correlation of reference markers, and finally, computer reconstruction of 2D-images enables the generation of a 3D-morphology of the dispersed nanoclays in the polymer matrix. Around 100 images from more than 300 x–y cross-sections are used for 3D-reconstructions. 3D-reconstructed images at different planes of PBSANC3 samples (after tensile tests) are shown in Fig. 4.13 [28]. The first clear observation is that the dispersed nanoclay platelets are oriented along the z-axis,

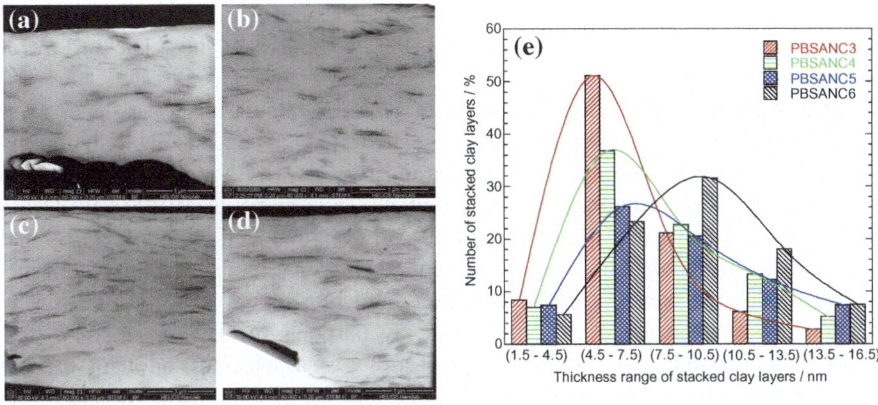

Fig. 4.12 The bright-field scanning transmission electron microscopy (STEM) images of four different nanocomposite systems, in which black entities represent the dispersed silicate layers: **a** PBSANC3 (containing 3 wt% C30B), **b** PBSANC4 (containing 4 wt% C30B), **c** PBSANC5 (containing 5 wt% C30B), and **d** PBSANC6 (containing 6 wt% C30B). **e** The number of stacked silicate layers (in %) for the different nanocomposites is plotted against the thickness of the stacked silicate layers (in nm) determined on the basis of the STEM images. Image J software was used to analyze STEM images (for each sample 280 tactoids were considered). Reproduced with permission from [24]. Copyright 2010, Elsevier Science Ltd

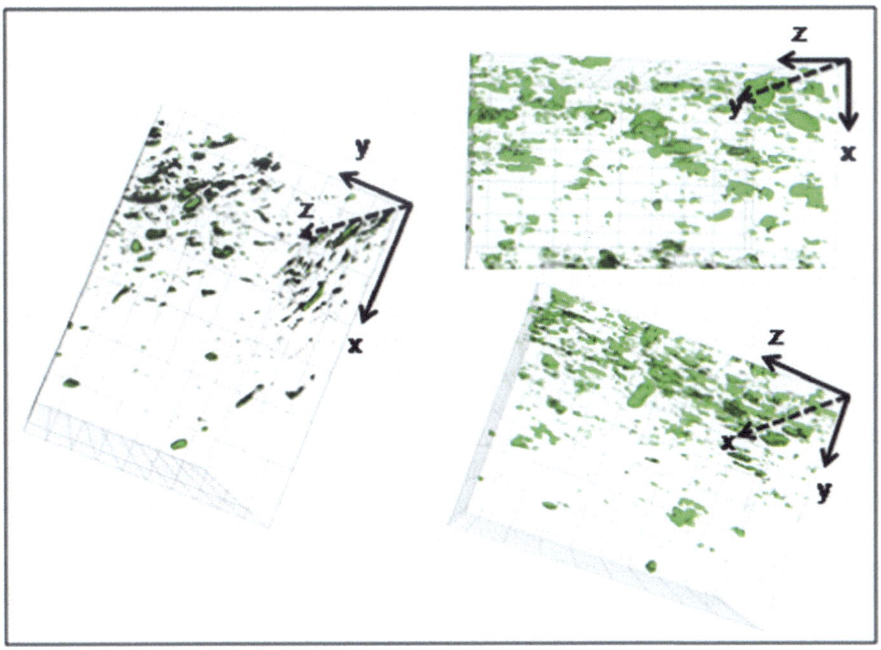

Fig. 4.13 The 3D-constructed images at different planes of the PBSANC sample by focused ion beam scanning electron microscope

i.e., the direction of elongation during the tensile test. The second observation is that the thicknesses of the clay layers are in the xy- and yz-planes. The surface of the clay platelets is in the xz-plane.

3D distributions of nanoclay platelets in polymer blends have also been visualized using electron tomography (or 3D TEM) [55]. Using a Gatan Ultrascan camera (resolution of 2–5 nm), 2D projection images with tilt angles ranging from −45° to +45° were automatically acquired in 2° increments at 200 kV. Using TEMography™ software, the tilted images were aligned to a common origin and subsequently reconstructed using a simultaneous iterative reconstruction technique. 3D distributions of nanoclay platelets in the blends are shown in Fig. 4.14a–d. The dark features represent the dispersed nanoclay particles in the blend composites. Figure 4.14a′–d′ and a″–d″ show 3D projections of a selected region of the 3D TEM images (refer to Fig. 4.14a–d). The blue markers in Fig. 4.14a′–d′ and a″–d″ represent the dispersed nanoclay particles and the silver, gold, and yellow colors in the background represent nylon 6 (N6), ethyl vinyl alcohol (EVOH), and the N6/EVOH blend, respectively. The images of various samples clearly show that the intercalated silicate layers of the nanoclay were preferentially oriented. It was clear that the nanoclay was located only in the interphase region in N6/EVOH/MB and formed core–shell particles. Such localization of intercalated silicate layers could suppress coalescence and stabilize the blend morphology. However, the intercalated silicate layers were well dispersed in the blend matrix of the N6/EVOH/OMMT composite; it was difficult to differentiate between the two phases.

4.3.2 Fourier Transformed Infrared Spectroscopy

FTIR spectroscopy is a powerful and versatile tool for analyzing the structure of multicomponent systems such as polymer blends, nanocomposites, and blend nanocomposites at the molecular level. The presence of additives, contaminants, degradation by-products, and chemical bonding at the interface of multi-layered polymer systems, polymer blends and/or the polymer–nanoparticle interface can be resolved using this technique [56]. The IR measurements in transmission mode require optically thin samples, preferably with thickness in the micron range. Various analysis modes can be used. A line scan across the cross-section can provide detailed information regarding the composition of the layers of multi-layered structure or laminates. This technique can also be employed to probe the absorption topography of polymer nanocomposites, as shown in Fig. 4.15 [57]. A noticeable change in the topography can be observed after changing the nanoclays (C20A and C30B) in the PBSA nanocomposites. The different nanoclays had different dispersion and interactions with the PBSA matrix, resulting in different topographical features.

In another study, FTIR spectra were extracted at different positions of absorbance images, as shown in Fig. 4.16. The absorption topography of PA and a PA PNC are shown in images (a) and (a′). False color absorbance images are shown

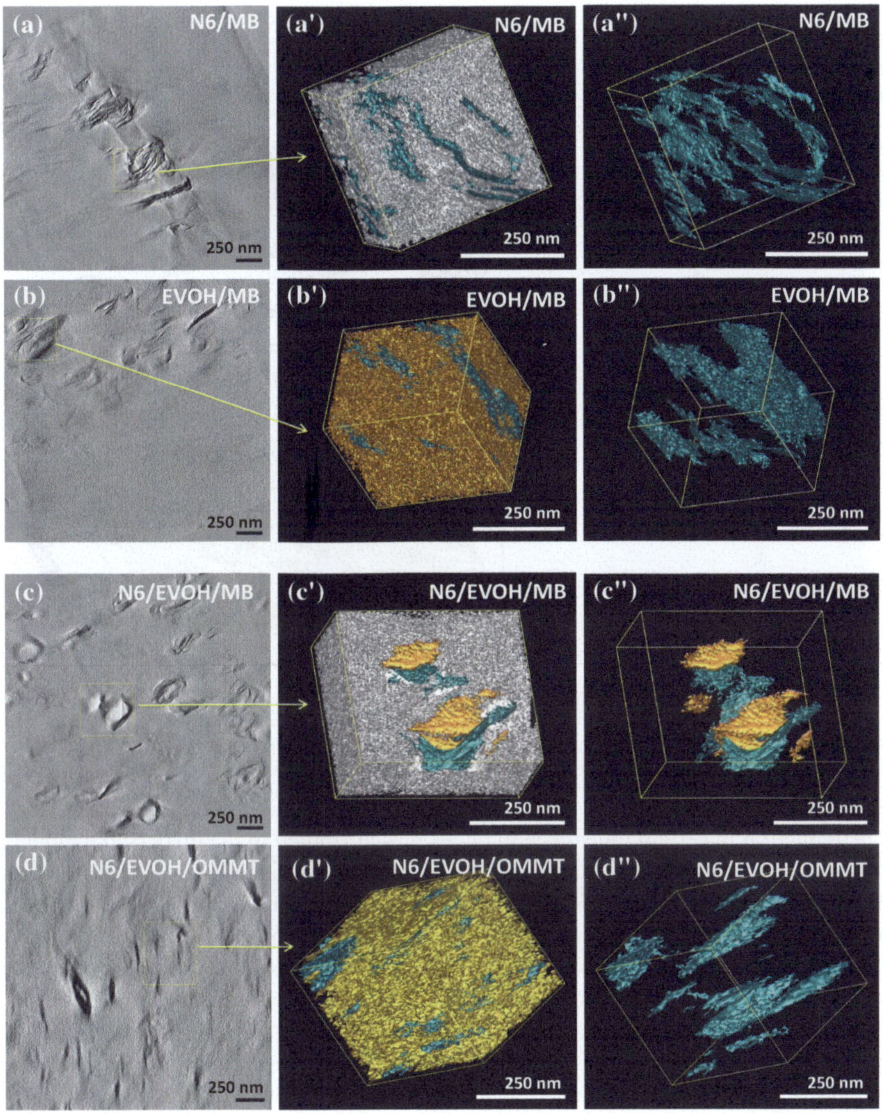

Fig. 4.14 **a–d** the 3D distributions of nanoclay platelets in the blends. The black markers represent the dispersed nanoclay particles in the blend composites. **a′–d′** and **a″–d″** the 3D projection of a selective region of 3D TEM images. Reproduced with permission from [55]. Copyright 2017, Elsevier Science Ltd

in (b) and (b′). The FTIR spectra at different positions (as indicated on the absorbance images) are shown in parts (c) and (c′). It can be seen that the absorbance was quite uniform for the PA film and it increased for the PA PNC (see Fig. 4.16). The FTIR spectra collected at two different positions (labeled 10 and 12) on the PA

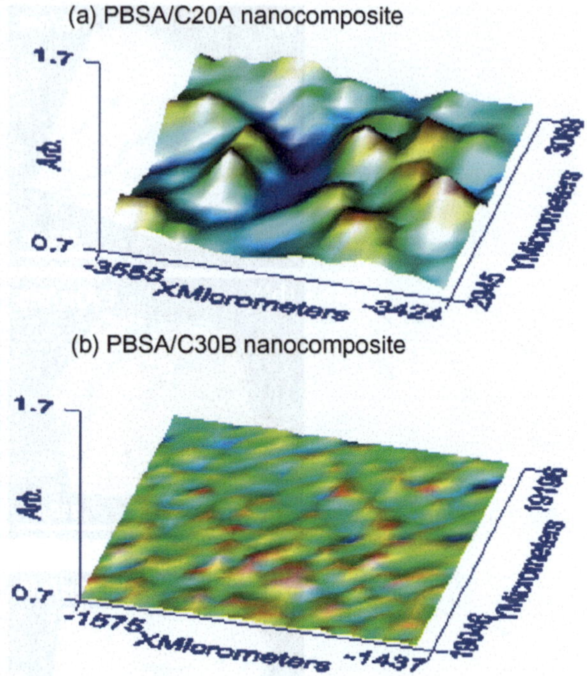

Fig. 4.15 Absorbance topography of **a** PBSA/C20A and **b** PBSA/C30B nanocomposites from FTIR microscopy. Reproduced with permission from [57]. Copyright 2008, Elsevier Science Ltd

film overlapped, indicating the same chemical composition (Fig. 4.16c). The results were different for PA PNC, which showed a change in the transmittance (or absorbance) when the spectra were collected at positions 13 and 15 (see Fig. 4.16c′). The reduction in certain transmission peaks was attributed to the dispersed nanoclay (Betsopa™ OM) platelets in the PA matrix. The characteristic PA peaks appeared at 3260, 2920, 1631, and 1536 cm^{-1} represented NH stretching, CH_2 stretching, amide I, and the amide II bands, respectively. The peak at 1036 confirmed the presence of the Si–O bond of the nanoclay in the PA PNC [58]. Since the spectra collected at different positions shown in Fig. 4.16b′ only differed in their intensity, it can be inferred that the nanoclays were well distributed throughout the matrix polymer. However, variations in the absorbance over the topography (Fig. 4.16a′) indicated the presence of stacked nanoclays along with well-dispersed ones.

4.4 Interfacial Analysis

Commonly used techniques, such as XRD and TEM, provide information about the inter-layer spacing of the nanoclay, but they cannot provide direct evidence on the molecular structure or the dynamics at the interface between the NP and polymer. Therefore, EPR, NMR, and dielectric spectroscopy techniques are used to investigate the interfacial regions, as discussed in this section.

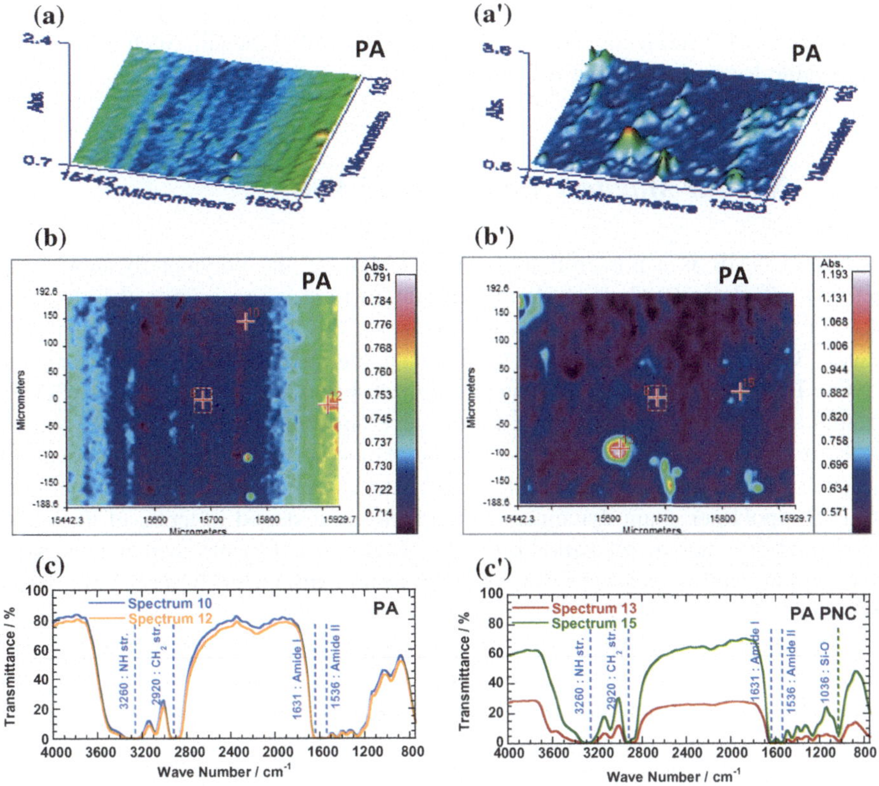

Fig. 4.16 **a**, **a′** Absorbance topography, **b**, **b′** False colour absorbance images, **c**, **c′** FTIR spectra of PA and PA PNC at different positions indicated in the absorbance images

4.4.1 Electron Paramagnetic Resonance (EPR)

It is well known that the interaction between a polymer and organically modified nanoclays plays an important role in the performance of the polymer nanocomposite. EPR techniques can be used to understand the structure and dynamics of the polymer–nanoclay interface by analyzing the surfactant layer. However, such measurements require spin labeling of the surfactant prior to organic modification of the nanoclay and use of a pre-deuterated polymer [59]. Steric acid and nitroxide molecules are commonly used for spin labeling of the surfactant, while catamine is used for spin labeling of the nanoclay surface [60]. Depending on the trans/gauche conformation, surfactant chains exhibit temperature-dependent heterogeneous mobility, as well as different dynamics along the alkyl chain [59]. These dynamics change in the PNC when polymer chains intercalate into nanoclay galleries. Electron spin echo envelope modulation (ESEEM) spectroscopy can be used to determine the anchoring position of the polymer and surfactant. For instance,

Schleidt et al. [59] observed that deuterated polystyrene anchors were quite close to the middle of the surfactant chain, instead of at the end of the tail. Such information is valuable for understanding the properties of the nanocomposites.

4.4.2 Nuclear Magnetic Resonance (NMR)

Low-field NMR is a powerful tool for investigating the polymer dynamics and network effects in PNCs [61]. As the relaxation time is sensitive to the chemical structure, interaction process, and homogeneity, the spin lattice relaxation time (e.g., $^{13}CT_1$, $^{1}H\ T_1$) can be used to elucidate the structure and dynamics of the PNCs. The dispersed nanoclay platelets restrict the mobility of the polymer chain around the nanoclay particles [62]. Latest developments in NMR technology have enabled interpretation of the effect of the nanoclay on the chain dynamics of polymers in the PNCs under external mechanical stress and at high temperature [63]. The polymer chain dynamics can be better understood in terms of the transverse nuclear magnetic relaxation time (T_2). T_2 is affected by physical and chemical cross-linking and is measured using a Hahn Echo experiment. Böhme and Scheler [63] noticed that in a PP nanocomposite polymer–filler interaction imposed an overall restriction on the motion that extended beyond the polymer–polymer interaction. The mobile and rigid components of T_2 can be extracted from the transverse magnetization relaxation function $M(t)$ as follows [64].

$$M(t) = M_0 \exp\left[-\frac{t}{T_2^{mobile}}\right] + (1 - M_0) \exp\left[-\frac{t}{T_2^{rigid}}\right] \tag{4.47}$$

where, M_0 is the fraction of mobile chains outside the adsorption layer, and T_2^{mobile} and T_2^{rigid} are the long and short spin–spin relaxation times, respectively.

4.4.3 Dielectric Spectroscopy

Dielectric spectroscopy yields information about the molecular motion and relaxation process in a polymeric material subjected to an electrical field. It provides some insight into the effects of networks formed by the dispersed NPs in the matrix polymer [65]. Two major polarization mechanisms studied by dielectric spectroscopy are polarization due to orientation of dipoles and charge migrations. These responses are important for the design of optoelectronic switches, printed board circuit, fuel cells, and other devices.

In the case of PNC, interfacial interaction between the polymer chains and nanofiller (weak but numerous bonding sites) can create long-range repulsive forces between particles. Such repulsive forces influence the polarization and separation of charges [66]. The dielectric permittivity characterizes the degree of electrical polarization in the material under the influence of an external electric field. Nelson and Hu [65] noticed that a nanocomposite exhibited a higher relative permittivity as a function of frequency than its microcomposite counterpart. In addition, the change in permittivity during curing can indicate the extent of cross-linking. By knowing the permittivity, the impedance (and hence, resistance) can be calculated. The extent of reaction or degree of curing (α) can be estimated as follows [67]:

$$\frac{\alpha}{\alpha_m} = \frac{\log(\rho) - \log(\rho_0)}{\log(\rho_m) - \log(\rho_0)} \tag{4.48}$$

Here, α_m is the maximum extent of reaction, ρ_m is the corresponding resistivity, and ρ_0 and ρ are the resistivity at the initial condition and when the degree of curing reaches α, respectively. The dielectric constant (i.e., ratio of the permittivity of a material to that of air/free space) and loss as a function of frequency during isothermal curing at a particular temperature can be used to describe the dynamics related to the curing process.

The dielectric constant depends on the properties of the interfacial region in the nanocomposite [68]. For example, long chain fluoro polymer (shell) attached to the $BaTiO_2$ NP (core) results compact surface and ordered structure. As a result the molecular chain mobility of fluoro polymer chains on $BaTiO_2$ reduces. Therefore, this kind of core-shell structure of NP creates loose interfacial region when dispersed in other polymer matrix. On the other hand, short chains attached to the nanoparticle allow high molecular chain mobility and disordered structures, leading to compact interfaces in the nanocomposite. As a consequence of the compact interface, the dielectric constant increases and the dielectric loss decreases. High interfacial polarization and low dielectric loss are highly desirable for electronic and electrical applications. However, it is very challenging to fulfil such criteria in PNCs unless polarized interfacial charges effectively increase the local electrical field in the polymer matrix along the applied electric field [69]. This is possible when NPs are interconnected by forming chains, clusters, or even aggregates. Dielectric spectroscopy can be employed to investigate the segmental dynamics of multi-layered films [70]. When interpenetration of the consecutive polymer layers is negligible, the interface does not have a significant effect on the dynamics. Therefore, dielectric spectroscopy is an indirect method for interfacial analysis.

4.5 Physical Effects: Rheological, Mechanical, and Barrier Properties

The physical properties of a PNC, such as its flow behavior (rheological properties), stiffness and toughness (mechanical properties), and gas barrier performance can also be used to evaluate the structure and dispersion of NPs in the polymer matrix. The properties of the polymer depend on its inherent structure (e.g., linear, branched, or crosslinked) and crystallinity and the composition (e.g., mono-polymer or co-polymer) [71]. The shape, aspect ratio, and physical interaction between the polymer and nanofillers in the nanocomposite can produce long range connectivity or the formation of network structures [71, 72]. Any transformation in the internal structure results in changes in the flow behavior and viscoelastic properties of the nanocomposites. For instance, Fig. 4.17 shows changes in the storage/elastic (G′) and loss/viscous (G″) moduli with changes in the nanoclay content in a PBSA matrix during melt-state rheological tests. It can be seen that for PBSANC4 at high frequency, G′ dominates over G″, while frequency below 1 rad/s G″ started to dominate [73]. Such behavior is quite common as the internal structures of the nanocomposite are rigid during rapid motion (high frequency oscillations). In this state, more deformation energy can be stored and the loss of deformation energy by friction between polymer chains due to their relative motion is reduced. Thus, the elastic behavior shows increasing dominance with increasing frequency, while with decreasing frequency, the network of entanglements has sufficient time to start disentangling and hence, nanocomposite samples show increasing flexibility and mobility. At very low frequency, most of the deformation energy is lost by frictional heating effects between polymer chains due to their mutual relative motion. Such a trend was observed for changes in G′ and G″ with increasing nanoclay loading in PBSANC5 and PBSANC6. In PBSANC5, G′ and G″ superimposes on each other over a range of frequency; while in PBSANC6, G′ dominates over G″ over the entire frequency range examined. For a closely packed network system (e.g., the presence of a percolation network in a flocculated nanocomposite) minimal deformation is possible and hence, solid-like behavior dominates in PBSANC5. A further increase in the nanoclay loading (PBSANC6) restricts the mobility of polymer chains; hence, G′ dominates over G″ over the entire frequency range. Therefore, such changes in the viscoelastic properties with nanoclay loading indicate changes in the dispersion characteristics in PBSANCs. The SAXS and STEM analyses discussed above verified this interpretation of the rheological behavior for the various PBSANCs.

Usually G′ curves are used to analyze the structural strength or consistency. The effect of different kinds of structures can be distinguished, particularly in the low frequency region of frequency sweeps. Ideally, macromolecules of polymers are assumed to be linear. For such unlinked polymers, the viscoelastic properties can be explained by Maxwell's model (a combination of spring and dashpots in series). The frequency dependence of G′ and G″ can then be obtained as follows:

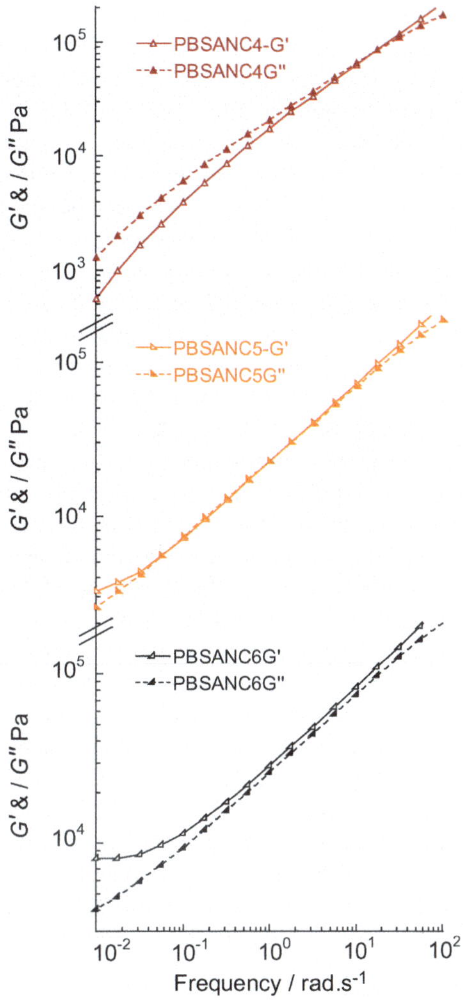

Fig. 4.17 Structural modifications observed in nanocomposites with increase in nanoclay (C30B) content. The PBSA nanocomposites with various wt% of C30B such as 4, 5, 6, were correspondingly abbreviated as PBSANC4, PBSANC5, and PBSANC6

$$G' = G_P \cdot \frac{\omega^2 \lambda^2}{\left(1 + \omega^2 \lambda^2\right)} \tag{4.49}$$

$$G'' = G_P \cdot \frac{\omega \lambda}{\left(1 + \omega^2 \lambda^2\right)} \tag{4.50}$$

In a nanocomposite, the viscoelastic behavior of the polymer changes from liquid-like (i.e., G′ $\alpha\ \omega^2$ and G″ $\alpha\ \omega$ from 4.49 to 4.50) to solid-like (i.e., G′ and G″ $\alpha\ \omega^0$) [74]. In Fig. 4.17, PBSANC4 shows liquid-like behavior. The solid-like behavior arises from the formation of network superstructures in PBSANC5 and then in PBSANC6 [74, 75].

PNCs exhibit enhanced structural strength, mainly due to an increased number of entanglements/network structures; the entanglements can be physical or chemical. Chemo-rheology involves the time- and temperature-dependent chemical bond formation and/or curing. Park and Jana [76] demonstrated that the elastic force exerted by cross-linked structures facilitates the dispersion of nanoclays in an epoxy matrix. It was observed that the gel temperature and time both decreased in the presence of NPs (e.g., CNTs, nanoclay) [77, 78]. Structural decomposition and regeneration (thixotropy and rheopexy) are critically important for self-healing of the polymer and nanocomposite [79, 80]. A thixotropic material exhibits a reduction in structural strength under shear force and regains the structural strength while at rest. The opposite trend is called the rheopexy. The shearing used for this purpose is high shear (above the linear viscoelastic, LVE, region) where the resting condition is simulated by the behavior in the LVE region.

Variations in the zero shear viscosity as a function of nanoclay loading can provide crucial information regarding the morphology of the blend nanocomposite. Ojijo et al. [81] elucidated the effect of nanoclay concentration on the morphology development in PLA and PBSA blends. As shown in Fig. 4.18, the authors observed a sudden increase in zero shear viscosity above 2 wt% nanoclay content. At concentrations ≤ 2 wt%, the nanoclay resided within the PBSA phase, and increased the viscosity of PLA phase only slightly. Above 2 wt% concentration, the nanoclay played an important role in controlling the zero-shear viscosity. Therefore, a disruptive change in blend morphology is expected at this point.

In addition to the flow behavior, the tensile properties, particularly the strength and elongation at break, indicated that the morphology of blend nanocomposites changed above 2 wt% nanoclay (B2; see Fig. 4.19). The authors correlated the morphology obtained from SEM analysis with the mechanical and the flow properties. Pure PLA has a much higher modulus than pure PBSA and hence, the modulus of the blend decreased significantly compared to pure PLA. However, the modulus of the composites tended to increase with increasing clay content, peaking at around 6 wt% due to the reinforcing effect of the nanoclay. However, the trends

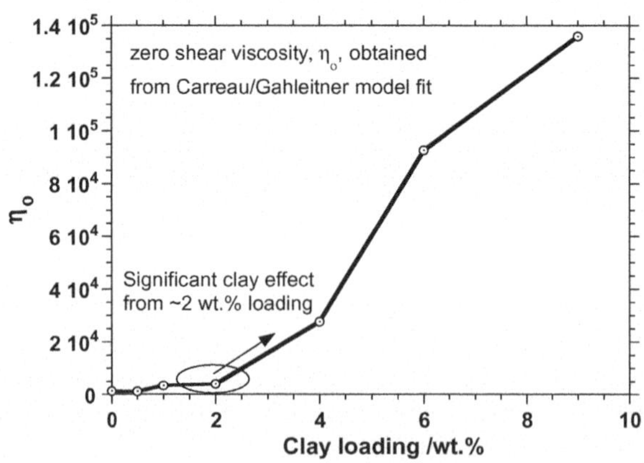

Fig. 4.18 Zero-shear viscosity as a function of nanoclay concentration in the PLA/PBSA blend. Reproduced with permission from [81]. Copyright 2012, the American Chemical Society

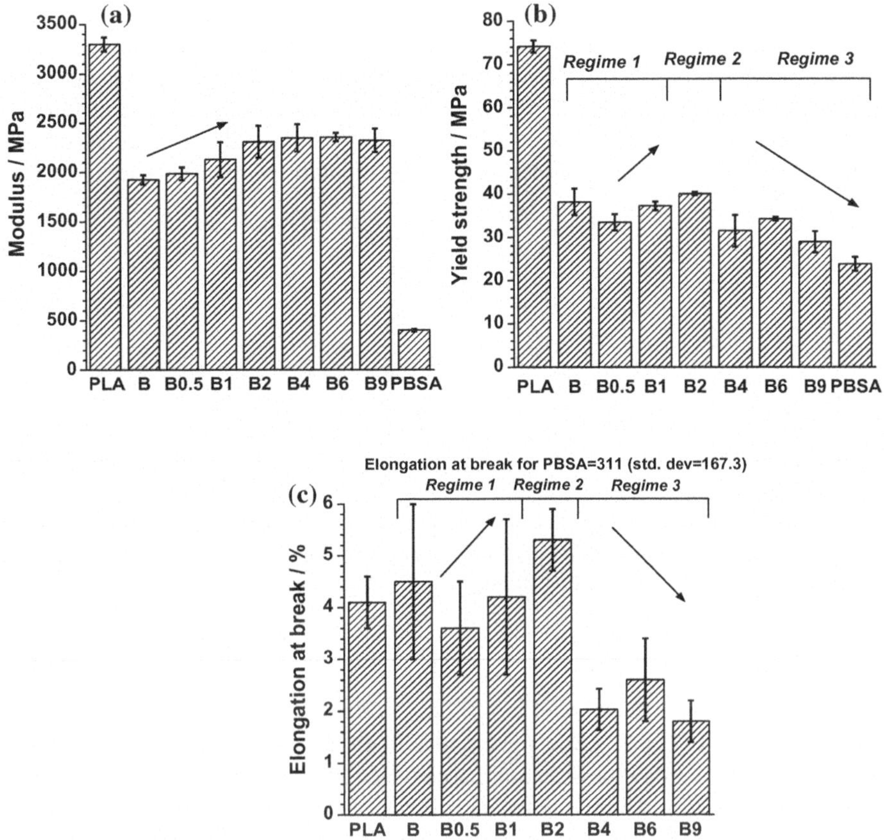

Fig. 4.19 Tensile properties of compression molded and annealed (80 °C for 12 h) samples of neat PLA, PBSA, blend (B) and blend nanocomposites (B0.5, B1, B2, B4, B6, and B9, each having 0.5, 1, 2, 4, 6, and 9 wt% of nanoclay, respectively). Reported values are averages of six independent experiments with standard deviations as error bars. The elongation at break of PBSA is 311% (std. dev. = 167.3). Reproduced with permission from [81]. Copyright 2012, the American Chemical Society

in the yield strength and elongation at break were confined to different regimes. Poor adhesion between PLA and PBSA results in a reduction in strength and elongation in the blend. A 0.5 wt% loading of nanoclay (in B0.5) seemed insufficient to affect the compatibility of the blend polymer. Increasing the nanoclay loading to 1 wt% (B1) served two purposes: stiffening the PBSA, and enhancing stress transfer between the PLA and PBSA phases due to the common adsorption of the two polymers on the clay surface. In B2 (regime 2), the nanoclays resided, not only within the PBSA phase and at the interface, but also within PLA. Hence, further improvements in the strength and elongation at break in B2 were observed.

The drastic reduction in strength and elongation at break in regime 3 of the composites was attributed to agglomeration of nanoclays in the blend matrices.

It has been shown, that in polymer nanocomposites, the interphases can occupy a significant volume fraction and influence the mechanical properties [82]. Although the Halpin-Tsai model (4.51) is commonly used to predict the mechanical performance of nanocomposites, it underpredicts the modulus in many cases.

$$E_R = \frac{1 + \eta\xi\varphi_f}{1 - \eta\varphi_f} \tag{4.51}$$

$$\eta = \left(\frac{E_f}{E_m} - 1\right) / \left(\frac{E_f}{E_m} + \xi\right) \tag{4.52}$$

Here, E_R is the ratio of the Young's modulus of the nanocomposite (E_c) to that of the matrix polymer (E_m), φ_f and E_f represent the volume fraction and modulus of the NPs, respectively.

ξ is the reinforcing factor which depends on the fiber geometry, packing geometry, and loading conditions. For a rectangular fiber cross-section with a length l and diameter d in a hexagonal array,

$$\xi = 2\left(\frac{l}{d}\right) \tag{4.53}$$

For spherical NPs, $\xi = 2$, while for circular fibers in a square array, $\xi = 1$. For a rectangular fiber with a cross-sectional area of length a and width b in a hexagonal array, $\xi = \sqrt{3}\ \ln\,(a/b)$, where a is the direction of loading. A recent study proposed that the volume fraction of the interphase (φ_i) for spherical NPs can be given by (4.54) [82].

$$\varphi_i = \left[\left(\frac{R + R_i}{R}\right)^3 - 1\right]\varphi_f \tag{4.54}$$

Here, R and R_i are the radius of NPs and the interphase thickness, respectively. By adding the interphase effect, the Halpin-Tsai model for spherical particles reduces to:

$$E_R = \frac{1 + 2\eta\varphi_f + 2\eta_i\left[\left(\frac{R+R_i}{R}\right)^3 - 1\right]\varphi_f}{1 - \eta\varphi_f - \eta_i\left[\left(\frac{R+R_i}{R}\right)^3 - 1\right]\varphi_f} \tag{4.55}$$

where,

$$\eta_f = \left(\frac{E_f}{E_m} - 1\right) / \left(\frac{E_f}{E_m} + 2\right) \tag{4.56}$$

and

$$\eta_i = \left(\frac{E_i}{E_m} - 1\right) / \left(\frac{E_f}{E_m} + 2\right) \tag{4.57}$$

Here, E_i is the Young's modulus of the interphase. Equation (4.55) accurately predicted the Young's modulus of the nanocomposites (containing spherical particles like silica, alumina, and calcium carbonate) assuming the role of the interphase [82].

It is well known that nanoclays dispersed in polymer nanocomposites impede permeating gas molecules by creating a tortuous path [83–87]. Therefore, any reduction in the gas permeation can be correlated with the extent of delamination and distribution of the nanoclays in the PNCs. The permeability of the gas molecule is estimated as the product of solubility and diffusion. Solubility is the partitioning behavior of a permeate molecule between the surface of the polymer and the surrounding headspace. Diffusion is the rate of movement of permeate molecule through the polymer matrix. According to Fick's law, the permeate flux (J) can be determined by (4.58).

$$J = -D\frac{\Delta c}{l} \tag{4.58}$$

where, D, Δc, and l represent the diffusion coefficient, change in permeate concentration, and thickness of the film, respectively.

According to Henry's law:

$$J = -D.S\frac{\Delta p}{l} \tag{4.59}$$

where, S and Δp are the ratio of equilibrium concentration of dissolved permeate to its partial pressure (c/p) and the pressure difference across the film, respectively. Combining Fick's law and Henry's law, the permeability can be written as:

$$P = D.S = \frac{J.l}{\Delta p} \tag{4.60}$$

Therefore, P is directly proportional to the permeate flux and the thickness of the film, and inversely proportional to the pressure difference across the film. D can further be correlated with the shape and size of the permeate molecule (A), free volume (f), and the minimum hole size required for a diffusion jump (B) using (4.61).

$$D = A^{-\frac{B}{f}} \tag{4.61}$$

In polymers, the glass transition temperature (T_g) is an indicator of the change in free volume. Incorporation of NPs can affect T_g and hence the gas permeability. In addition to the above-mentioned parameters, the polymer crystals and chemical structure of the polymer also play an important role in controlling the gas barrier behavior of polymeric materials. Therefore, high-aspect-ratio nanofillers and their influence on the intrinsic properties of the matrix polymer are critical parameters for enhancing the gas barrier properties of the nanocomposites.

4.6 Conclusion

The techniques described in this chapter are complementary to each other. While microscopy allows direct visualization of dispersed nanoparticles over a small area, X-ray scattering analyzes the overall dispersion and distribution characteristics, degree of anisotropy in the system, and crystal structures. However, these methods cannot provide direct evidence on the molecular structure or the dynamics of the interface. Electron paramagnetic resonance, nuclear magnetic resonance, and dielectric spectroscopy allow investigation of the interfacial regions. Other indirect measurements for evaluating the structure of the nanocomposites include rheological (studying the flow behavior), mechanical, and gas diffusion tests.

Acknowledgements The authors would like to thank the Department of Science and Technology and the Council for Scientific and Industrial Research, South Africa, for financial support.

References

1. Drummy LF, Wang YC, Schoenmakers R, May K, Jackson M, Koerner H, Farmer BL, Mauryama B, Vaia RA. Morphology of layered silicate- (nanoclay-) polymer nanocomposites by electron tomography and small-angle X-ray scattering. Macromolecules. 2008;41:2135–43.
2. Ho DL, Briber RM, Glinka CJ. Characterization of organically modified clays using scattering and microscopy techniques. Chem Mater. 2001;13:1923–31.
3. Mittelbach R, Glatter O. Direct structure analysis of small-angle scattering data from polydisperse colloidal particles. J Appl Crystallogr. 1998;31:600–8.
4. Schnablegger H, Singh Y. A practical guide to small angle X-ray scattering. Austria: Anton Par GmbH; 2006.
5. Glatter O, Kratky O. Small angle X-ray scattering. London: Academic Press; 1982. ISBN 0-12-286280-5.
6. Glatter O. Data evaluation in small angle scattering: calculation of the radial electron density distribution by means of indirect Fourier transformation. Acta Phys Austriaca. 1977;47:83–102.
7. Glatter O. A new method for the evaluation of small-angle scattering data. J Appl Crystallogr. 1977;10:415–21.

8. Bergmann A, Fritz G, Glatter O. Solving the generalized indirect Fourier transformation (GIFT) by Boltzmann simplex simulated annealing (BSSA). J Appl Crystallogr. 2000;33:1212–6.
9. Brunner-Popela J, Glatter O. Small-angle scattering of interacting particles. I. Basic principles of a global evaluation technique. J Appl Crystallogr. 1997;30:431–42.
10. Weyerich B, Brunner-Popela J, Glatter O. Small-angle scattering of interacting particles. II. Generalized indirect Fourier transformation under consideration of the effective structure factor for polydisperse systems. J Appl Crystallogr. 1999;32:197–209.
11. Hosemann R, Bagchi SN. Direct analysis of diffraction by matter. Amsterdam, The Netherlands: North-Holland; 1962.
12. Guinier A. X-ray diffraction in crystals, imperfect crystals and amorphous bodies. Ontario, Canada: General Publishing Company; 1994.
13. Caillé A, Seances CR. Remarks on the scattering of X-rays by A-type smectics. Actes Soc Hist B. 1972;274:891–3.
14. Zhang R, Tristram-Nagle S, Sun W, Headrick RL, Irving TC, Suter RM, Nagle JF. Small-angle X-ray scattering from lipid bilayers is well described by modified Caillé theory but not by paracrystalline theory. Biophys J. 1996;70:349–57.
15. Zhang R, Suter RM, Nagle JF. Theory of the structure factor of lipid bilayers. Phys Rev E. 1994;50:5047–60.
16. Frühwirth T, Fritz G, Freiberger N, Glatter O. Structure and order in lamellar phases determined by small-angle scattering. J Appl Crystallogr. 2004;37:703–10.
17. Glatter O. Convolution square root of band-limited symmetrical functions and its application to small-angle scattering data. J Appl Crystallogr. 1981;14:101–8.
18. Glatter O, Hainisch B. Improvements in real-space deconvolution of small-angle scattering data. J Appl Crystallogr. 1984;17:435–41.
19. Glatter O. Comparison of two different methods for direct structure analysis from small-angle scattering data. J Appl Crystallogr. 1988;21:886–90.
20. Li TC, Ma J, Wang M, Tjiu C, Liu T, Huang W. Effect of clay addition on the morphology and thermal behaviour of polyamide 6. J Appl Polym Sci. 2007;103:1191–9.
21. Ganguli A, Bhowmick AK. Insights into montmorillonite nanoclay based ex situ nanocomposites from SEBS by small angle X-ray scattering and modulated DSC studies. Macromolecules. 2008;41:6246–53.
22. Preschilla N, Sivalingam G, Rasheed AS, Tyagi S. Quantification of organoclay dispersion and lamellar morphology in poly(propylene) nanocomposites with small angle X-ray scattering. Polymer. 2008;49:4285–97.
23. Nawani P, Burger C, Chu B, Hsiao BS, Tsou AH, Weng W. Characterization of nanoclay orientation in polymer nanocomposite film. Polymer. 2010;51:5255–66.
24. Bandyopadhyay J, Ray SS. The quantitative analysis of nano-clay dispersion in polymer nanocomposites by small angle X-ray scattering combined with electron microscopy. Polymer. 2010;51:1437–49.
25. Carli LN, Bianchi O, Machado G, Crespo JS, Mauler RS. Morphological and structural characterization of PHBV/organoclay nanocomposites by small angle X-ray scattering. Mater Sci Eng. 2013;33:932–7.
26. Yadav R, Naebe M, Wang X, Kandasubramanian B. Structural and thermal stability of polycarbonate decorated fumed silica nanocomposite via thermomechanical analysis and in-situ temperature assisted SAXS. Sci Rep. 2017;7:7706. https://doi.org/10.1038/s41598-017-08122-7.
27. Jouault N, Dalmas F, Boúe F, Jestin J. Multiscale characterization of filler dispersion and origins of mechanical reinforcement in model nanocomposites. Polymer. 2012;53:761–75.
28. Bandyopadhyay J, Malwela T, Ray SS. Study of change in dispersion and orientation of clay platelets in a polymer nanocomposite during tensile test by variostage small-angle X-ray scattering. Polymer. 2012;53:1747–59.

29. Bandyopadhyay J, Ray SS. Determination of structural changes of dispersed clay platelets in a polymer blend during solid-state rheological property measurement by small-angle X-ray scattering. Polymer. 2011;52:2628–42.
30. Gurun B, Bucknall DG, Thio YS, Teoh CC, Harkin-Jones E. Multiaxial deformation of polyethylene and polyethylene/clay nanocomposites: in situ synchrotron small angle and wide angle X-ray scattering study. J Polym Sci Part B Polym Phys. 2011;49:669–77.
31. Nishada T, Obayashi A, Haraguchi K, Shibayama M. Stress relaxation and hysteresis of nanocomposite gel investigated by SAXS and SANS measurement. Polymer. 2012;53:4533–8.
32. Bandyopadhyay J, Sinha Ray S, Scriba M, Wesley-Smity J. A combined experimental and theoretical approach to establish the relationship between shear force and clay platelet delamination in melt-processed polypropylene nanocomposites. Polymer. 2014;55:2233–45.
33. Pujari S, Dougherty L, Mobuchon C, Carreau PJ, Heuzey M-C, Burghardt WR. X-ray scattering measurements of particle orientation in a sheared polymer/clay dispersion. Rheol Acta. 2011;50:3–16.
34. Thompson A, Bianchi O, Amorim CLG, Lemos C, Teixeira SR, Samios D, Giacomelli C, Crespo JS, Machado G. Uniaxial compression and stretching deformation of an i-PP/EPDM/organoclay nanocomposite. Polymer. 2011;52:1037–44.
35. Yamashita M, Kato M. Lamellar crystal thickness transition of melt crystallized isotactic polybutene-1 observed by small-angle X-ray scattering. J Appl Crystallogr. 2007;40:s650–5.
36. Fu Q, Heck B, Strobl G, Thoman Y. A temperature- and molar mass-dependent change in the crystallization mechanism of poly(1-butene): transition from chain-folded to chain-extended crystallization? Macromolecules. 2001;34:2502–11.
37. Marega C, Causin V, Saini R, Marigo A. A direct SAXS determination of specific surface area of clay in polymer-layered silicate nanocomposites. J Phys Chem B. 2012;116:7596–602.
38. Bunge HJ. Influence of texture on powder diffraction. Text Microstruct. 1997;29:1–26.
39. Courgneau C, Domenek S, Lebossé R, Guinault A, Avérous L, Ducruet V. Effect of crystallization on barrier properties of formulated polylactide. Polym Int. 2012;62:180–9.
40. Incarnato L, Scarfato P, Russo GM, Maio LD, Iannelli P, Acierno D. Preparation and characterization of new melt compounded copolyamide nanocomposites. Polymer. 2003;44:4625–34.
41. Zhu W, Chen T, Li Y, Lei J, Chen X, Yao W, Duan T. High performances of artificial nacre-like graphene oxide-carrageenan bio-nanocomposite films. Materials. 2017;10:536. https://doi.org/10.3390/ma10050536.
42. Hikku GS, Jeyasubramanian K, Venugopal A, Ghosh R. Corrosion resistance behaviour of graphene/polyvinyl alcohol nanocomposite coating for aluminium-2219 alloy. J Alloy Compd. 2017;716:259–69.
43. Maravi S, Bajpai J, Bajpai AK. Improving mechanical and electrical properties of poly(vinyl alcohol-*g*-acrylic acid) nanocomposite films by reinforcement of thermally reduced graphene oxide. Polym Sci Ser A. 2017;59:751–63.
44. Di Mauro A, Cantarella M, Nicotra G, Pellegrino G, GulinonA Brundo MV, Privitera V, Impellizzeri G. Novel synthesis of ZnO/PMMA nanocomposites for photocatalytic application. Sci Rep. 2017;7:40895. https://doi.org/10.1038/srep40895.
45. Shanthala VS, Devi SN, Murugendrappa MV. Synthesis, characterization and DC conductivity studies of polypyrrole/copper zinc iron oxide nanocomposites. J Asian Ceram Soc. 2017;5:227–34.
46. Vaez M, Alijini S, Omidkhah M, Moghaddam AZ. Synthesis, characterization and optimization of N-TiO_2/PANI nanocomposite for photodegradation of acid dye under visible light. Polym Compos. 2017. https://doi.org/10.1002/pc.24574.
47. Chen Y-H, Zhong GJ, Wang Y, Li ZM, Li L. Unusual tuning of mechanical properties of isotactic polypropylene using counteraction of shear flow and β-nucleating agent on β-form nucleation. Macromolecules. 2009;42:4343–8.
48. Zhang Y-F, Chang Y, Li X, Xie D. Nucleation effects of a novel nucleating agent bicyclic [2,2,1]heptane di-carboxylate in isotactic polypropylene. J Macromol Sci. 2011;50:266–74.

49. De Santis F, Pantani R. Optical properties of polypropylene upon recycling. Sci World J. 2013;2013:1–7.
50. Malas A, Bharati A, Verkinderen O, Goderis B, Moldenaers P, Cardinaels R. Effect of the GO reduction method on the dielectric properties, electrical conductivity and crystalline behavior of PEO/rGO nanocomposites. Polymers. 2017;9:613. https://doi.org/10.3390/polym9110613.
51. Nwofe PA, Ramakrishna Reddy KT, Sreedevi G, Tan JK, Forbes I, Miles RW. Single phase, large grain, p-Conductivity-type SnS layers produced using the thermal evaporation method. Energy Proc. 2012;15:354–60.
52. Wang Y, Tang W, Zhang L. Crystalline size effects on texture coefficient, electrical and optical properties of sputter-deposited Ga-doped ZnO thin films. J Mater Sci Technol. 2015;31:175–81.
53. Ilican S, Caglar M, Caglar Y. Determination of the thickness and optical constants of transparent indium-doped ZnO thin films by the envelope method. Mater Sci Pol. 2007;25:709–18.
54. Bandyopadhyay J, Sinha Ray S. Mechanism of enhanced tenacity in a polymer nanocomposite studied by small-angle X-ray scattering and electron microscopy. Polymer. 2010;51:4860–6.
55. Bandyopadhyay J, Sinha Ray S, Salehiyan R, Ojijo V. Effect of the mode of nanoclay inclusion on morphology development and rheological properties of nylon6/ethyl-vinyl-alcohol blend composites. Polymer. 2017;126:96–108.
56. Bhargava R, Wang S-Q, Koenig JL. FTIR microscopy of the polymeric systems. Adv Polym Sci. 2003;163:137–91.
57. Ray SS, Bandyopadhyay J, Bousmina M. Influence of degree of intercalation on the crystal growth kinetics of poly[(butylene succinate)-co-adipate] nanocomposites. Eur Polymer J. 2008;44:3133–3145.
58. Alabarse FG, Conceição RV, Balzaretti NM. In-situ FTIR analyses of bentonite under high-pressure. Appl Clay Sci. 2011;51:202–8.
59. Schleidt S, Spiess HW, Jeschke G. A site-directed spin-labeling study of surfactants in polymer–clay nanocomposites. Colloid Polym Sci. 2006;284:1211–9.
60. Kielmann U, Jeschke G, García-Rubio G. Structural characterization of polymer-clay nanocomposites prepared by co-precipitation using EPR techniques. Materials. 2014;7:1384–408.
61. Papon A, Saalwächter K, Schäler K, Guy L, Montes H. Low-field NMR investigations of nanocomposites: polymer dynamics and network effects. Macromolecule. 2011;44:913–22.
62. da Silva E, Tavares MIB, Nogueira JS. Solid state evaluation of natural resin/clay nanocomposites. J Nano Res. 2008;4:117–26.
63. Böhme U, Scheler U. Interfaces in polymer nanocomposites—an NMR study. Proceedings of PPS-31. AIP Conf Proc. 2016;1713:090009-1–3.
64. Dewimille L, Bresson B, Bokobza L. Synthesis, structure and morphology of poly (dimethylsiloxane) networks filled with in situ generated silica particles. Polymer. 2005;46:4135–43.
65. Nelson JK, Hu Y. Nanocomposite dielectrics—properties and implications. J Phys D Appl Phys. 2005;38:213–22.
66. Lewis TJ. Interfaces: nanometric dielectrics. J Phys D Appl Phys. 2005;38:202–12.
67. Kenny JM, Trivisano A. Isothermal and dynamic reaction kinetics of high performance epoxy matrices. Polym Eng Sci. 1991;31:1426–33.
68. Wang K, Huang X, Huang Y, Xie L, Jiang P. Fluoro-polymer@$BaTiO_3$ hybrid nanoparticles prepared via RAFT polymerization: Toward ferroelectric polymer nanocomposites with high dielectric constant and low dielectric loss for energy storage application. Chem Mater. 2013;25:2327–38.
69. Zhang G, Brannum D, Dong D, Tang L, Allahyarov E, Tang S, Kodweis K, Lee J-K, Zhu L. Interfacial polarization-induced loss Mechanisms in polypropylene/$BaTiO_3$ nanocomposite dielectrics. Chem Mater. 2016;28:4646–60.

70. Casalini R, Prevosto D, Labardi M, Roland CM. Effect of interface interaction on the segmental dynamics of poly(vinyl acetate) investigated by local dielectric spectroscopy. ACS Macro Lett. 2015;4:1022–6.
71. Abraham J, Sharika T, George SC, Thomas S. Rheological percolation in thermoplastic polymer nanocomposites. Rheol Open Access. 2017;1:1–15.
72. Knauret ST, Douglas JF, Starr FW. The effect of nanoparticle shape on polymer-nanocomposite rheology and tensile strength. J Polym Sci Part B Polym Phys. 2007;45: 1882–97.
73. Bandyopadhyay J, Ray SS, Maiti A, Khatua B. Thermal and rheological properties of biodegradable poly[(butylene succinate)-co-adipate] nanocomposites. J Nanosci Nanotechnol. 2010;10:4184–95.
74. Krishnamoorti R, Yurekli K. Rheology of polymer layered silicate nanocomposites. Curr Opin Colloid Interface. 2001;6:464–70.
75. Eslami H, Grmela M, Bousmina M. A mesoscopic tube model of polymer/layered silicate nanocomposites. Rheol Acta. 2009;48:317–31.
76. Park JH, Jana SC. Mechanism of exfoliation of nanoclay particles in epoxy-clay nanocomposites. Macromolecules. 2003;36:2758–68.
77. Terenzi A, Vedova C, Leilli G, Mijovic J, Torre L, Valentini L, Kenny JM. Chemorheological behaviour of double-walled carbon nanotube-epoxy nanocomposites. Compos Sci Technol. 2008;68:1862–8.
78. Kim J-T, Martin D, Halley P, Kim DS. Chemorheological studies on a thermoset PU/clay nanocomposite system. Compos Interfaces. 2012;14:449–65.
79. Fox J, Wie J, Greenland B, Burattini S, Hayes W, Colquhoun H, Mackay M, Rowan S. High strength, healable, supramolecular polymer nanocomposites. J Am Chem Soc. 2012;134: 5362–8.
80. Wang Y, He J, Aktas S, Sukhishvilli SA, Kalyon DM. Rheological behaviour and self-healing of hydrogen-bonded complexes of a tribock Pluronic® copolymer with weak polyacid. J Rheol. 2017;61:1103. https://doi.org/10.1122/1.4997591.
81. Ojijo V, Ray SS, Sadiku R. Effect of nanoclay loading on the thermal and mechanical properties of biodegradable polylactide/poly[(butylene succinate)-co-adipate] blend composites. ACS Appl Mater Interfaces. 2012;4:2395–405.
82. Zare Y. Development of Halpin-Tsai model for polymer nanocomposites assuming interphase properties and nanofiller size. Polym Test. 2016;51:69–73.
83. Arunvisut S, Phummanee S, Somwangthanaroj A. Effect of clay on mechanical and gas barrier properties of blown film LDPE/clay nanocomposites. J Appl Polym Sci. 2007;106:2210–7.
84. Golebiewski J, Rozanski A, Dzwonkowski J, Galeski A. Low density polyethylene–montmorillonite nanocomposites for film blowing. Eur Polymer J. 2008;44:270–86.
85. Lotti C, Isaac CS, Branciforti MC, Alves RM, Liberman S, Bretas RE. Rheological, mechanical and transport properties of blown films of high density polyethylene nanocomposites. Eur Polymer J. 2008;44:1346–57.
86. Yeh J-T, Chang C-J, Tsai F-C, Chen K-N, Huang K-S. Oxygen barrier and blending properties of blown films of blends of modified polyamide and polyamide-6 clay mineral nanocomposites. Appl Clay Sci. 2009;45:1–7.
87. Garofalo E, Fariello ML, Di Maio L, Incarnato L. Effect of biaxial drawing on morphology and properties of copolyamide nanocomposites produced by film blowing. Eur Polymer J. 2013;49:80–9.

Chapter 5
Impact of Melt-Processing Strategy on Structural and Mechanical Properties: Clay-Containing Polypropylene Nanocomposites

Dimakatso Morajane, Suprakas Sinha Ray, Jayita Bandyopadhyay and Vincent Ojijo

Abstract Processing conditions (e.g., temperature profile, feed point, screw speed, feed rate, and screw element configuration) and how nanocomposites are prepared in the extruder have a vital effect on the dispersion of nanoclay. The resultant morphology of nanocomposites is not only a question of shear stress or residence time, but also a result of the entire mechanical and thermal history of the material when extruded. Hence, this study intends to extensively investigate the aspects of processing conditions, such as temperature profile, feed point, screw speed, feed rate, and screw element configuration, and the relationship between the different parameters (optimal conditions). The clay-containing polymer nanocomposite has been selected as a model system and the effects of nanoclay and maleic anhydride grafted PP loading on the dispersion of nanoclay in the PP nanocomposite have been investigated. The aim of this study is to investigate ways of improving the dispersion of nanoclay in the PP matrix and to determine how the state of dispersion affects the morphology and properties of resultant PP nanocomposites. A co-rotating twin-screw extruder was used to produce nanocomposites owing to the flexibility of the screw profile, screw speed, feed rate, and material feeding in different areas of the machine.

D. Morajane · S. Sinha Ray (✉) · J. Bandyopadhyay · V. Ojijo
DST-CSIR National Centre for Nanostructured Materials,
Council for Scientific and Industrial Research, Pretoria 0001, South Africa
e-mail: rsuprakas@csir.co.za

D. Morajane · S. Sinha Ray
Department of Applied Chemistry, University of Johannesburg,
Doornfontein 2028, Johannesburg, South Africa
e-mail: dmorajane@csir.co.za; ssinharay@uj.ac.za

J. Bandyopadhyay
e-mail: jbandyopadhyay@csir.co.za

V. Ojijo
e-mail: vojijo@csir.co.za

S. Sinha Ray (ed.), *Processing of Polymer-based Nanocomposites*,
Springer Series in Materials Science 277,
https://doi.org/10.1007/978-3-319-97779-9_5

5.1 Introduction

This chapter focuses on the impact of melt-processing strategy on structural and mechanical properties of clay-containing polypropylene (PP) nanocomposites. The first phase addresses the first objective: To evaluate the effect/impact of processing strategy, such as feed point of the nanoclay and processing sequence type (i.e. MB, masterbatch and SP, single pass methods) on the dispersion of nanoclay in the PP matrix. In doing this, different processing sequences were investigated and their effect on nanoclay dispersion was established. The second phase addresses the second specific objective of this work: To examine the screw element configuration design and its role in dispersion of nanoclay in PP. By doing this, the effects of several screw element configuration designs on nanoclay dispersion were investigated. The third phase addresses the third specific objective of this study: To study the influence of processing conditions, such as temperature profile during the compounding of PP nanocomposites, on the dispersion of nanoclay. Through this, the optimal processing conditions were established. The fourth phase addresses the fourth specific objective of this study: To study the influence of maleic anhydride grafted PP (PP-g-MA) content during compounding of PP nanocomposites on the dispersion of nanoclay. Through this, the optimal PP-g-MA content was established. The fifth phase addresses the fifth specific objective of this work: To investigate the effects of silicate concentration, feed rate, and screw speed on residence time. In so doing, the ideal feed rate and screw speed were established.

5.2 Materials and Methods

PP homopolymer (trade name PP HHR 102) is a commercial product purchased from Sasol, South Africa. According to the supplier, it has a molecular weight (M_w) of 200 kg/mol, density of 0.905 g/cm^3, and melt flow index of 2.0 g/10 min (at 230 °C and 2.16 kg).

PP-g-MA was used as a compatibilizer. It was obtained from Vin Poly Additives, India. According to the supplier, the grafting level of MA in the PP-g-MA is 1%.

The organically-modified nanoclay used was Betsopa OM™ (Betsopa), a commercially available bentonite from our laboratory. Betsopa is a South African calcium bentonite modified with dimethyl dehydrogenated tallow quaternary ammonium surfactant. Throughout manuscript it has been abbreviated as GF290.

Ultra-blue was used as a tracer with a mass of 8.99 mg per pellet. It was donated to the CSIR Nano Center as a sample from Masterbatch SA (Pty) Ltd.

Prior to extrusion, PP-g-MA and GF290 were dried in a vacuum oven for 24 h at 80 °C for all the phases. TE-30 co-rotating TSE from Nanjing Extrusion Machinery Co. Ltd., which has screw diameter of 30 mm and L/D of 40, was used to produce the nanocomposites. The process of TSE starts by feeding the material into the

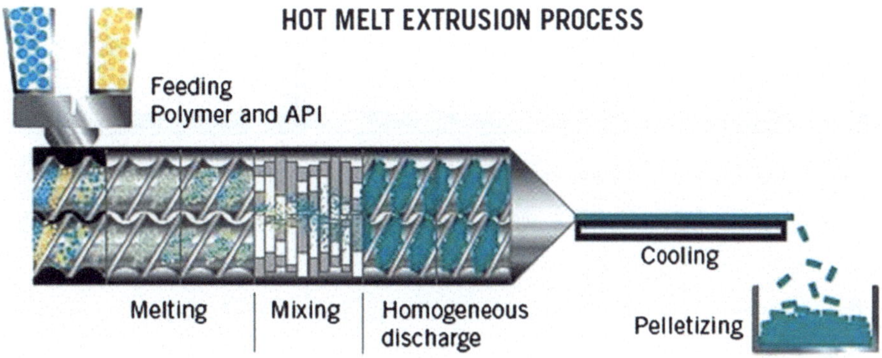

Fig. 5.1 Schematic diagram of hot melt extrusion process [1]. Adapted from http://www.particlesciences.com/docs/technical_briefs/TB_2011_3.pdf

hopper, melting, homogeneous discharge of the material, cooling, and finally palletizing, as shown in Fig. 5.1. The TSE is a flexible process; hence, this particular study configured the screws in different ways, allowing the degree of mixing and conveying of materials in the extruder to be controlled, as demonstrated in Fig. 5.2. Different screw element configuration designs were labelled SC1, SC2, and SC3.

5.3 Processing Strategies and Their Effect on Nanocomposite Structural and Mechanical Properties

5.3.1 Protocol 1: Effect of the Processing Sequences

In the first phase, the PP nanocomposites were prepared through two main processes: (i) a double-pass method that entails the preparation of an MB followed by its dilution, and (ii) an SP method. In the double-pass method, an MB (inorganic content determined from thermogravimetric analysis ~31 wt%) was prepared in the first pass, and then diluted in the same TSE in the second pass to achieve an inorganic nanoclay content of 3 wt%. On the other hand, the SP method involved the direct preparation of composites with the final intended concentrations in just a single pass. This involves compounding all the components at once to achieve an inorganic nanoclay content of 3 wt%. Each of these two processes had different sequences of nanocomposite production, as illustrated in Fig. 5.3.

The composition of all the nanocomposites is the same, where 95 wt% of PP is blended with 2 wt% of PP-g-MA and 3 wt% of nanoclay. The only difference is how they were prepared in the extruder. A brief explanation of the sequence is as follows:

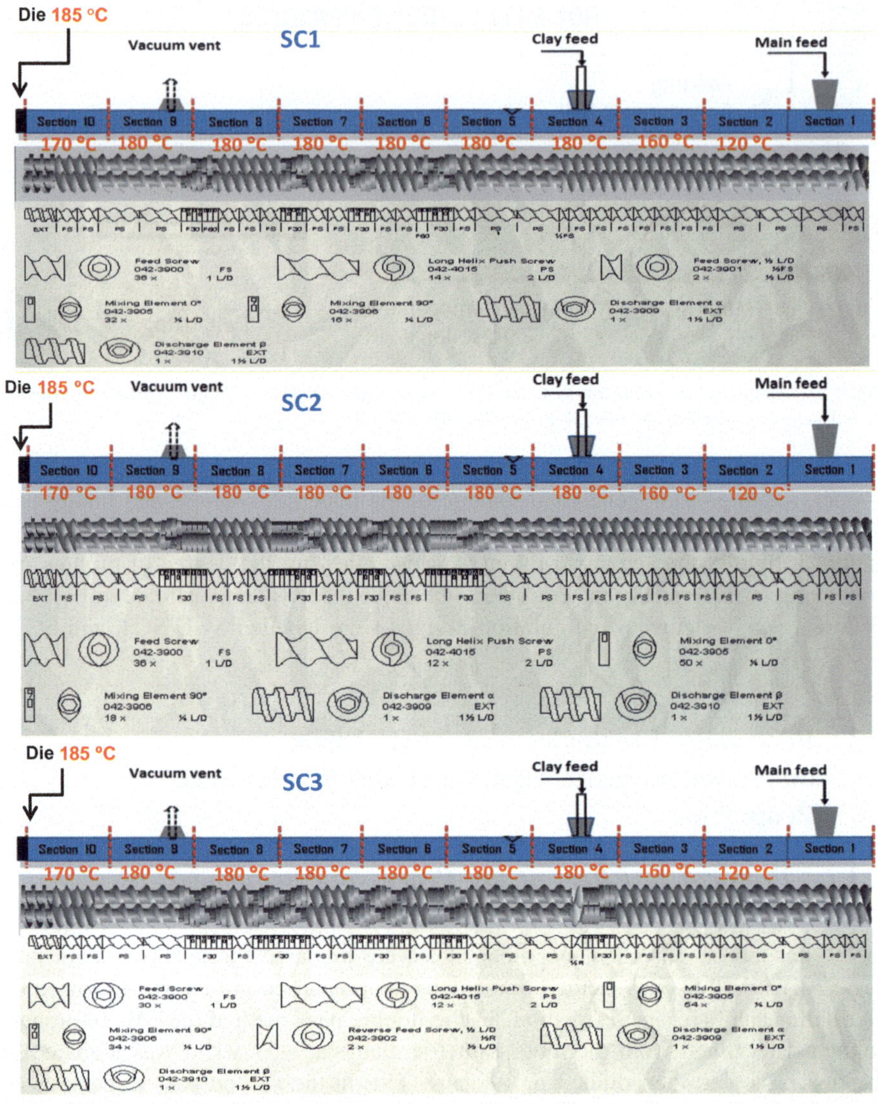

Fig. 5.2 Screw element configuration showing different elements of the screw and the different zone temperatures. SC1: 16% knedding elements; SC2: 21% knedding elements; SC3: 29% knedding

- *Sequence 1*: PP-g-MA/MB of ~31 wt% inorganic content was prepared by feeding PP-g-MA into the main feeder, followed by nanoclay inclusion at zone 4. The prepared MB was then be diluted in PP in such a way that the final nanocomposite has 3 wt% inorganic content. The nanocomposite is coded as (95/2/3 S1).

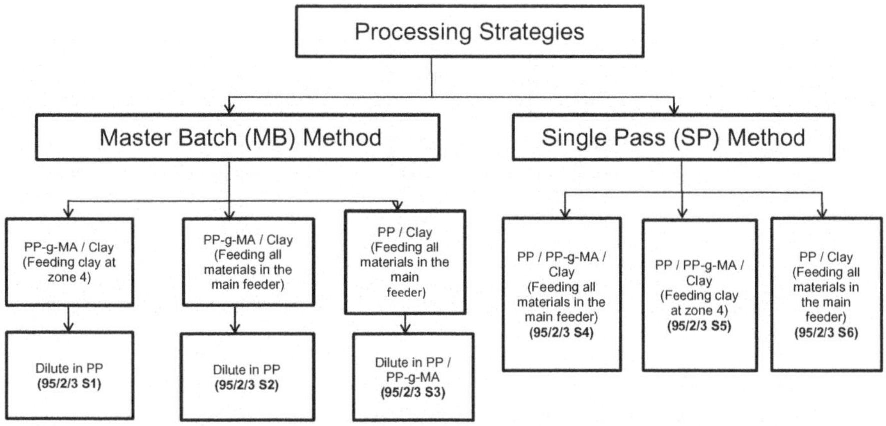

Fig. 5.3 Flowchart of six different mixing sequences of the produced PP nanocomposites

- *Sequence 2*: PP-g-MA and nanoclay were dry-mixed and fed through the main feeder. The prepared MB was then diluted in PP to get 3 wt% inorganic nanoclay content in the nanocomposite. The nanocomposite is coded as (95/2/3 S2).
- *Sequence 3*: Unlike sequence 2, in this case PP and nanoclay were dry-mixed and fed through the main feeder. The prepared MG was then diluted in PP/PP-g-MA. The final nanocomposite has 3 wt% inorganic nanoclay content. The nanocomposite is coded as (95/2/3 S3).
- *Sequence 4*: PP-g-MA/nanoclay/PP nanocomposite with 3 wt% inorganic nanoclay content was prepared by feeding all materials into the main feeder. The nanocomposite is coded as (95/2/3 S4).
- *Sequence 5*: PP-g-MA/nanoclay/PP PCNs of ~3 wt% inorganic nanoclay content was prepared by feeding PP-g-MA/PP through the main feeder, followed by the nanoclay at zone 4. The nanocomposite is coded as (95/2/3 S5).
- *Sequence 6*: PP/nanoclay PCNs of ~3 wt% inorganic nanoclay content was prepared by feeding PP/clay through the main feeder. The nanocomposite is coded as (95/2/3 S6).

The screw speed and feed rate were kept constant at 202 rpm and 6.6 kg/h, respectively. The nanocomposites were produced using the screw configuration SC1, which had 21% of its length consisting of kneading elements. The temperatures of different zones in the extruder barrel were as follows: 120/160/180/180/180/180/180/180 and 170 °C for zones 2–10, respectively, while the die was set at 185 °C.

The XRD patterns of the nanoclay and PP nanocomposites prepared via MB and SP methods are presented in Fig. 5.4. The figure shows that a diffraction peak appears at an angle of 2.5° in the nanoclay; connotating a *d*-spacing of 3.5 nm. The diffraction peak appears almost at the same position in all the PNCs except 97/0/3 S6. The *d*-spacing of different PNCs are listed in Table 5.1. The marginal change in

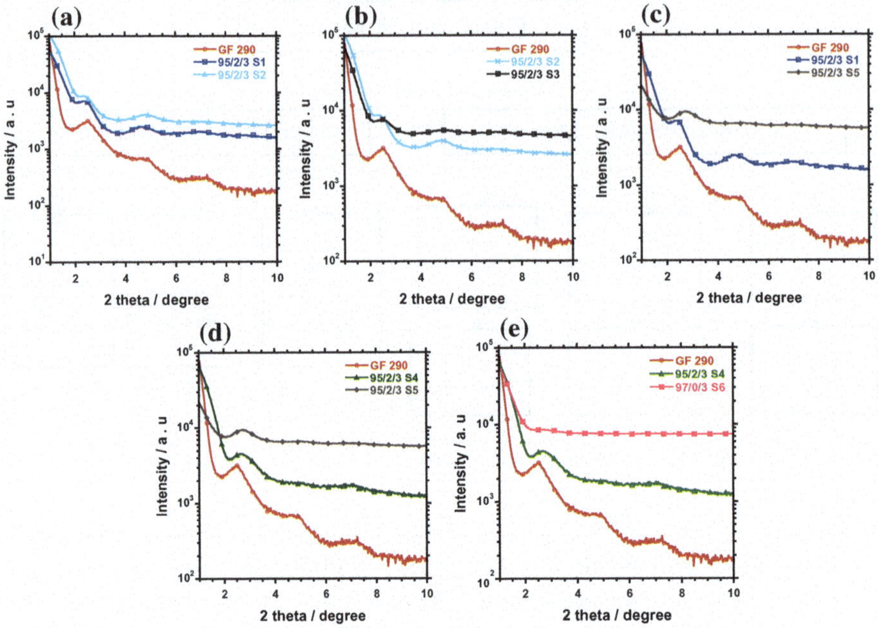

Fig. 5.4 X-ray diffractograms of the nanocomposites prepared via masterbatch (MB) and singple-pass (SP) methods: **a** effect of feeding nanoclay at different zones during the preparation of MB; **b** effect of PP-g-MA incorporation methods on the dispersion of nanoclay; **c** effect of processing methods (MB and SP) on the dispersion of nanoclay in the PP matrix, nanoclay was fed at zone 4 in both scenarios; **d** effect of feeding nanoclay at different zones in the extruder during SP method; and (e) effect of using compatibilizer in the SP method. 95/2/3 S6 is the control

Table 5.1 *d*-spacing of different PNC

Sample name PP/PP-g-MA/clay	2θ (°)	*d*-spacing (nm)
GF290	2.5	3.5
95/2/3 S1	2.5	3.5
95/2/3 S2	2.5	3.5
95/2/3 S3	2.6	3.4
95/2/3 S4	2.7	3.3
95/2/3 S5	2.7	3.3
97/0/3 S6	2.9	3.0

d-spacing indicates that the processing methods do not have a significant effect on the dispersion characteristics of nanoclay in the PP matrix. On the other hand, in the absence of PP-g-MA (in 95/0/3 S6), a shallow diffraction peak appears at an angle of approximately 3.0°. Therefore, *d*-spacing decreases slightly in 95/0/3 S6 when compared with other nanocompostes. It is noted that although PP-g-MA improves

Table 5.2 Melting and cooling profiles of neat PP and nanocomposites

Sample name	T_m (°C)	ΔH_c (J g^{-1})	T_c (°C)	% crystallity
Neat PP	164.3	91.6	116.2	45.6
95/2/3 S1	163.9	93.4	118.7	46.5
95/2/3 S2	164.3	99.4	120.9	49.5
95/2/3 S3	164.5	96.9	121.6	48.3
95/2/3 S4	165.7	110.9	123.4	55.2
95/2/3 S5	165.2	103.7	123.3	51.6
97/0/3 S6	165.0	100.9	123.5	50.3

the wetting characteristics of nanoclay in the PP matrix, the *d*-spacing of nanoclay remains unaltered in the nanocomposites containing PP-g-MA. It is well established that the dispersion of nanoclay in the polymer matrix depends on the surface peeling, followed by the diffusion of polymer chains in the nanoclay galleries [2]. Recently, Borse et al. [3] showed that while the peeling mechanism requires lower shear stresses that are achievable during melt extrusion, longer residence time is required for this mechanism to occur. In fact, the shear stresses required for peeling are significantly lower at higher peeling angles. Moreover, the peeling mechanism depends on the dimensions of the nanoclay platelets [2]. High shear stress during extrusion could break up the stacks into smaller ones and hinder the peeling process. Since the current screw profile contains 21% mixing elements, it is possible to achieve quite a high shear during the extrusion process. High shear might have prevented the peeling mechanism; hence, it can be expected that there will be a homogeneous distribution of small nanoclay tactoids in the nanocomposites, but not the high level of dispersion.

The TEM images in Fig. 5.5 show the outcome of processing strategy (MB and SP methods) on the dispersion and orientation of the nanoclay in the nanocomposites. The micrographs show that the nanoclay layers are better delaminated in samples prepared with MB method that initially had a compatibilizer (95/2/3 S1 and 95/2/3 S2), unlike 95/2/3 S3. However, less stacking of nanoclays is witnessed in 95/2/3 S2 than 95/2/3 S1; this observation agrees with the SAXS results. In addition, the distribution of nanoclay platelets was more ordered in 95/2/3 S2. This demonstrates the advantage of 95/2/3 S2, which gives nanoclay particles more residence time in the extruder barrel. The longer residence time enables better dispersion. Better dispersion enables easier orientation compared to 95/2/3 S1, where nanoclays are randomly oriented owing to less residence time. All samples prepared by SP method exhibit less dispersion of nanoclay platelets in the polymer matrix because of the presence of nanoclay tactoids, and the nanoclay is orientated randomly in the polymer matrix. The presence of compatibilizer in the 95/2/3 S4 and 95/2/3 S5 reduces the agglomeration of nanoclay in the polymer matrix by facilitating the miscibility of the nanoclay and PP.

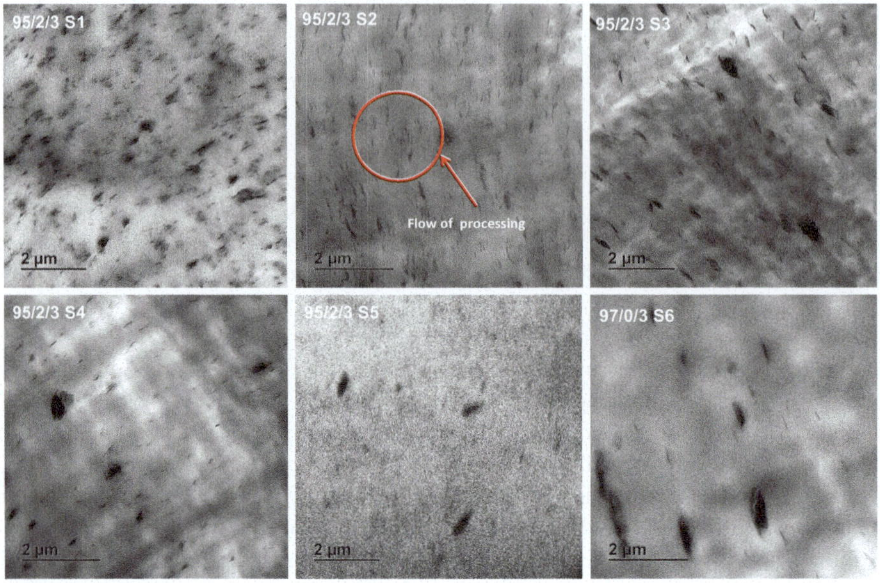

Fig. 5.5 TEM micrograms for nanocomposites prepared via masterbatch and single-pass methods at low magnification

Figure 5.6 shows the tensile modulus, elongation, and impact strength of neat PP and nanocomposites after the addition of 3 wt% inorganic nanoclay content. Figure 5.6a, b reveal that neat PP and nanocomposites exhibit modulus and elongation comparable to the nanocomposites produced by MB and SP methods. The feeding of nanoclay at different points of the extruder barrel also yielded composites with comparable moduli. This could be attributed to the less interfacial interaction between the nanoclay and PP matrix regardless of the processing strategy used to produce nanocomposites. However, the 95/2/3 S4 shows a higher modulus than nanocomposites owing to better crystallinity, as summarized in Fig. 5.2.

Figure 5.6c shows the impact strength of neat PP and nanocomposites. It is evident that the addition of nanoclay in the PP matrix decreases the impact strength of the nanocomposites from 61 to 26 kJ/m^2. This is attributed to the addition of the nanoclay that limits the molecular movement of the polymer chain, leading to the agglomeration of the nanoclay and poor interfacial interaction. The decrease in the impact strength from neat polymers to nanocomposites has been observed before. Forne et al. [4] previously reported the decrease in mechanical properties with different concentrations of nanoclay in a nylon system. Yuan and Misra [5] found that the addition of 4 wt% nanoclay in the HDPE system decreased the impact strength of the resultant nanocomposite. The difference in the impact strength between samples prepared via MB method and SP method is not significant. The impact strength of nanocomposites ranged from 26 to 36 kJ/m^2. Li et al. [6] previously reported the similar trend of the impact strength for MB and SP

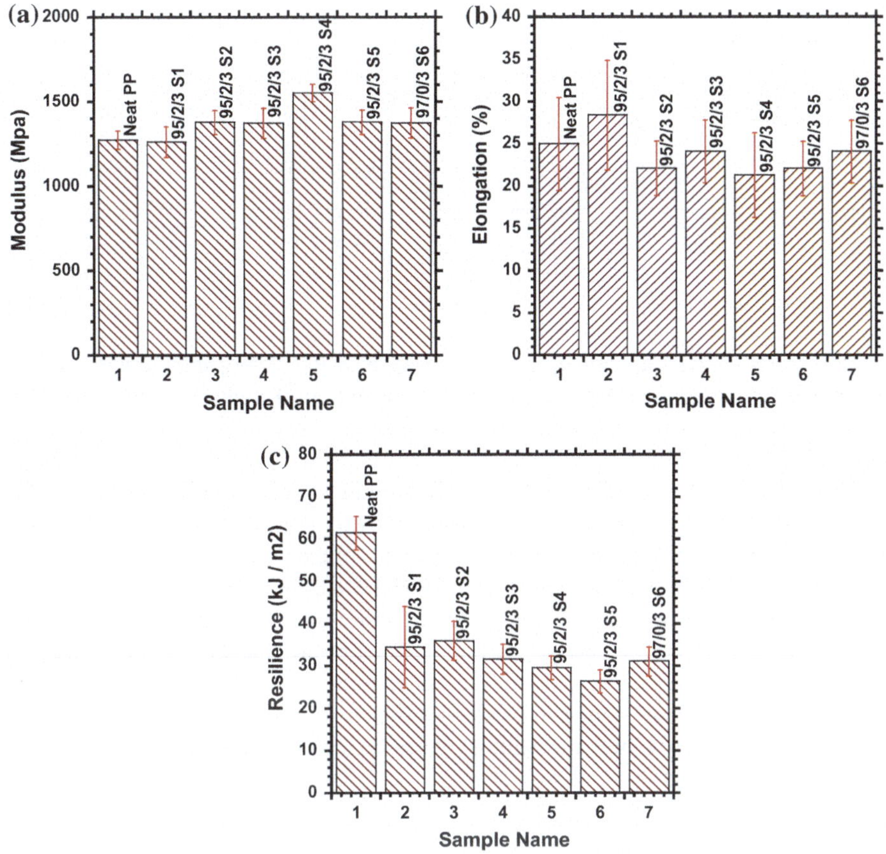

Fig. 5.6 **a** Tensile modulus, **b** elongation, and **c** impact strength at room temperature of neat PP and nanocomposites. The figure displays the average values obtained after ten tests were carried out

methods using the PP system. The authors revealed that there are two mechanisms affecting the impact fracture: initiation and propagation. They further revealed that the microvoid coalenscence in the microscale is not sensitive to the micro dispersion of organoclays in the PP matrix. For all samples, the enforcement level for the impact strength is quite similar.

Table 5.3 shows the HDT-Vicat of neat PP and nanocomposites prepared via MB and SP methods. Neat PP has a low HDT compared to the nanocomposites. The addition of the nanoclay enhanced the HDT of the nanocomposites by approximately 1–5 °C. However, the addition of the nanoclay in the PP matrix did not increase the HDT significantly; this result concurs with the degree of crystallinity results (refer to Table 5.2). The 95/2/3 S6 has a lower HDT than the nanocomposites produced owing to the poor dispersion as a result of the absence of the compatibilizer when preparing the sample. The HDT depends on the aspect

Table 5.3 HDT-Vicat data for various samples

Sample name PP/PP-g-MA/clay	HDT mean (°C)	Vicat mean (°C)
100/0/0 (neat PP)	50 ± 0.67	155 ± 0.3
95/2/3 S1	52 ± 1.41	154 ± 0.2
95/2/3 S2	54 ± 1.77	154 ± 0.3
95/2/3 S3	54 ± 1.77	155 ± 0.4
95/2/3 S4	55 ± 0.90	155 ± 0.3
95/2/3 S5	54 ± 1.14	155 ± 0.4
97/0/3 S6	51 ± 0.66	155 ± 0.2

ratio of the dispersed nanoclay particles because of the strong interaction between the matrix and silicate surface caused by the formation of hydrogen bonds [6]. The increase of HDT in the nanocomposites is vital not only for industry but also for molecular control of the silicate layers, which is crystallization through the interaction between PP molecules and SiO_4 tetrahedral layers in the MMT [7]. Nalini et al. [8] reported the increase in HDT from 99.87 to 103.39 °C using a PP system. This means that only 3.6% of HDT was achieved; in the current study, 10% of HDT was achieved. It is important to have high HDT because it will improve the capacity of the final product to withstand high temperatures; for example, these materials will not deform at high temperatures when placed in the microwave. The softening point of all materials is similar in all the nanocomposites and neat PP. This means that the softening point of the samples is not dependent on the addition of nanoclay to the polymer matrix and the strategy used to produce nanocomposites in the extruder.

5.3.2 *Protocol 2: Effect of the Screw Element Configuration Design*

In order to study the effect of the screw element configurations on the nanoclay dispersion in PP, two additional screw configurations were investigated, as shown in Diagram 5.2. For the high-shear screw configuration SC2, 29% of the length was comprised of kneading elements. On the other hand, the low-shear screw configuration SC3 had only 16% of the screw length consisting of kneading elements. The sequence that had better dispersion from previous experiments using the medium-shear screw configuration SC1 was chosen for further analysis, and then contrasted with those prepared using SC2 and SC3. The samples were coded as 95/2/3 SC1, 95/2/3 SC2, and 95/2/3 SC3. The PP nanocomposites using the other two screw configurations were prepared by mixing PP-g-MA and nanoclay in the main feeder to produce an MB of ~31 wt% inorganic nanoclay content, then diluted in PP to 3 wt% nanocomposites. The speed screw and feed rate were kept constant at 202 rpm and 6.6 kg/h, respectively. The temperatures of different zones in the

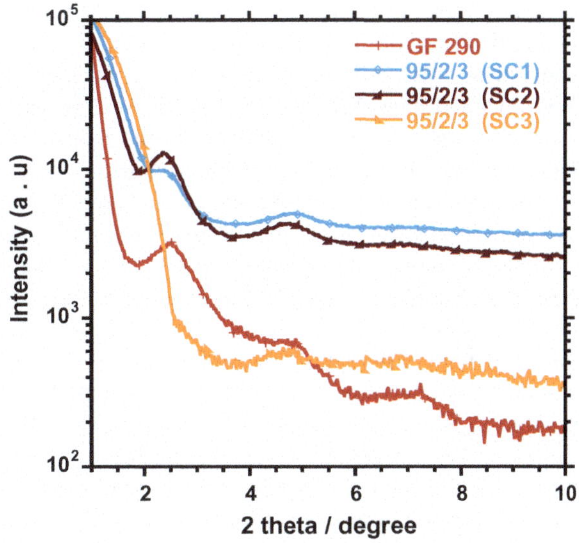

Fig. 5.7 X-ray diffractograms of the nanoclay and nanocomposites prepared via MB method using different percentages of screw element configuration design

extruder barrel were as follows: 120/160/180/180/180/180/180/180 and 170 °C for zones 2–10, respectively, while the die was set at 185 °C.

The XRD patterns of the virgin nanoclay together with nanocompoites prepared by MB method using different screw element configurations are shown in Fig. 5.7. The inclusion of the nanoclay in the PP matrix leads to a peak broadening of the 95/2/3 SC1 and 95/2/3 SC2 samples. Table 5.4 shows that the *d*-spacing(d_{001}) of the 95/2/3 SC1 and 95/2/3 SC2 are the same; however, the difference between these two samples is the intensity or the peak height. The 95/2/3 sample prepared with SC1 has low intensity owing to lower amount of nanoclay stacks in the sample, better morphology, and the peeling of the nanoclay. The 95/2/3 sample prepared by SC2, which consists of 29% mixing elements, has high intensity owing to the increase of nanoclay stacks present in the sample and the high shear used during extrusion. Lertwimolnun and Vergnes [8] concluded that the most severe screw profile is not necessarily efficient, indicating that other parameters have to be taken into account in the delamination mechanism. The evolution of dispersion along the screws is largely influenced by the feed rate. Furthermore, nanocomposite samples prepared using SC3 that consists of 16% mixing elements show that the nanoclay

Table 5.4 *d*-spacing of PNCs prepared with different screw configurations

Sample name PP/PP-g-MA/Clay	2θ (°)	*d*-spacing (nm)
GF290[a]	2.5	3.5
95/2/3 (SC1)	2.5	3.5
95/2/3 (SC2)	2.5	3.5
95/2/3 (SC3)	–	–

[a]Organically modified calcium bentonite, Betsopa OM™

platelets were not well-dispersed in the PP matrix. This is depicted by the non-appearance of the peak, which is due to the poor dispersion and distribution of nanoclays; this will be discussed later in the TEM section. Most probably, agglomerated nanoclays are localised in such a way that the indent X-ray did not interact properly with the nanoclay in the nanocomposite while the XRD experiment was conducted in the reflection mode.

The screw element configuration can change the morphology of nanocomposites owing to high shear forces that occur during processing (extrusion and/or injection) [9]. The influence of processing conditions on the morphology of PP nanocomposites was studied using TEM to evaluate the magnitude of nanoclay intercalation or exfoliation into the PP matrix. The TEM micrographs in Fig. 5.8 show samples prepared with different screw configurations in the extruder and the same concentration of nanoclay. Nanocomposites prepared with different shear stress display random orientation dispersion of the nanoclay in the 95/2/3 SC2 and 95/2/3 SC3 samples. The 95/2/3 SC1 sample produced better dispersion of the nanoclay. Generally, this suggests that the higher shear stress achieved in 95/2/3 SC1 enabled better dispersion compared to the low shear SC3 screw configuration. One possible conclusion is that a threshold shear is needed, beyond which no observable change is noted; in this case, it was SC1. Increasing the number of kneading blocks beyond that in SC1 did not improve the dispersion further. The 95/2/3 SC1 prepared with average shear of 21% mixing elements displayed elongated forms with better-separated nanoclay platelets and better distribution in the PP matrix. These results are supported by the SAXS (results are not reported here). The increased shear in the 95/2/3 SC2 was promoted by the addition of 8% mixing elements in the extruder, which assisted in breaking or the breakdown of the nanoclay platelets into bigger stacks than 95/2/3 SC1.

Figure 5.9 reveals the modulus, elongation, and impact strength of neat PP and nanocomposites. The impact strength of neat PP is higher than that of nanocomposites prepared with different screw element configurations in the extruder. The decrease in the impact strength of the nanocomposites is caused by the inclusion of 3 wt% inorganic nanoclay content in PP matrix. Using high shear in the extruder to

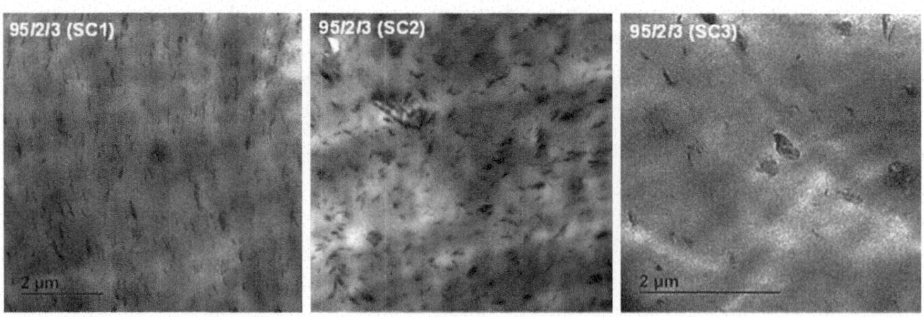

Fig. 5.8 TEM micrograms for nanocomposites prepared with MB method using different screw mixing elements in the extruder

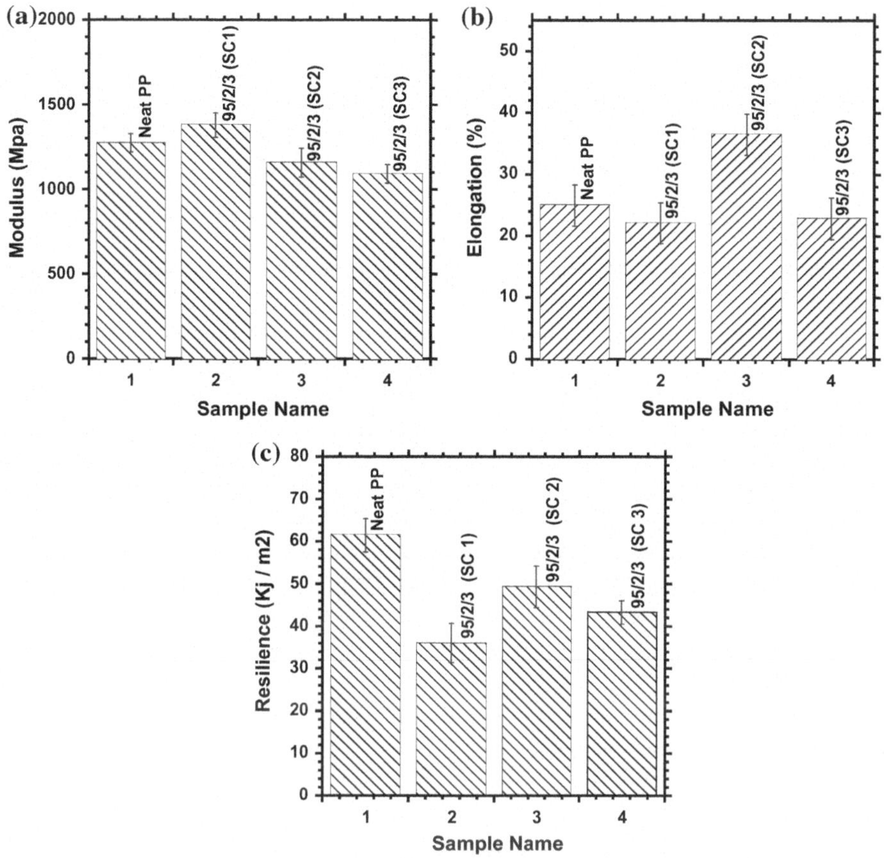

Fig. 5.9 **a** Tensile modulus, **b** elongation, and **c** impact strength at room temperature of neat PP and nanocomposites. The figure displays the average values obtained after ten tests were carried out using different screw mixing elements

produce 95/2/3 (SC2) nanocomposites led to better impact strength owing to the homogeneous dispersion-distribution of nanoclay. The 95/2/3 SC1 has lower impact strength because of the aligned orientation of the nanoclay, unlike the random orientation in 95/2/3 SC2 that led to high impact strength. Using high shear in the extruder increased the elongation and impact strength of the 95/2/3 SC2 owing to the well-distributed agglomerates that exist in the morphology of the PP matrix. The 95/2/3 SC1 has low impact strength due to the elongated structure in the PP matrix that served as a tension concentrator in the sample and reduced the impact strength compared to the 95/2/3 SC2 and 95/2/3 SC3. This study shows that well-distributed agglomerates of nanoclay leads to enhanced impact strength and Young's modulus. Santos et al. [10] revealed similar findings. The sample prepared with low shear (PPC20AP2) exhibited higher impact strength than the sample prepared with high shear (PPC20AP1) owing to the elliptical structures in the

Table 5.5 HDT-Vicat data for various samples

Sample name PP/PP-g-MA/clay	HDT mean (°C)	Vicat (°C)
100/0/0 (Neat PP)	50 ± 0.67	155 ± 0.3
95/2/3 (SC1)	54 ± 1.77	154 ± 0.3
95/2/3 (SC2)	66 ± 1.9	154 ± 0.1
95/2/3 (SC3)	68 ± 0.1	156 ± 0.5

PPC20AP1, which probably acted as tension concentrator by reducing the impact strength.

The 95/2/3 SC2 sample exhibited a higher modulus than other samples because of better dispersion of nanoclay platelets in the polymer matrix. Using 21% of mixing element in the extruder can increase the stiffness of the material. When comparing the different screw element configurations, the stiffness of the 95/2/3 SC2 and 95/2/3 SC3 were not significantly changed. A similar trend was observed for the Vicat of all three samples. This indicates that the different shear in the extruder does not affect the stiffness and Vicat of the nanocomposites.

Table 5.5 presents the HDT-Vicat of the nanocomposites and neat PP. Neat PP show low HDT compared to the nanocomposite. The decrease of HDT is caused by the inclusion of 3 wt% inorganic nanoclay content. When comparing the nanocomposites, it is observed that the 95/2/3 SC3 prepared with low shear display higher HDT than the 95/2/3 SC1 and 95/2/3 SC2 owing to the minor change in the crystallinity of the materials. For this study, SC1 was enough to break the nanoclay platelets in the polymer matrix.

5.3.3 *Protocol 3: Effect of Processing Conditions*

To study the effect of the processing conditions on the dispersion of nanoclays in PP, different temperature profiles listed in Table 5.6 were used. The PP nanocomposites were prepared by mixing PP-g-MA and nanoclay in the main feeder to produce an MB of ~31 wt% inorganic nanoclay content, before diluting in PP to 3 wt% nanocomposites. The nanocomposites were produced using SC1. The screw speed and feed rate were constant at 202 rpm and 6.6 kg/h, respectively.

Table 5.6 Different temperature profiles

Temperature profile (TP)	Different temperature profiles (°C)								
Barrel zones	2	3	4	5	6	7	8	9	10 (die)
95/2/3 180 °C	120	160	180	180	180	180	180	170	185
95/2/3 190 °C	120	160	190	190	190	190	190	170	185
95/2/3 200 °C	120	160	200	200	200	200	200	170	185

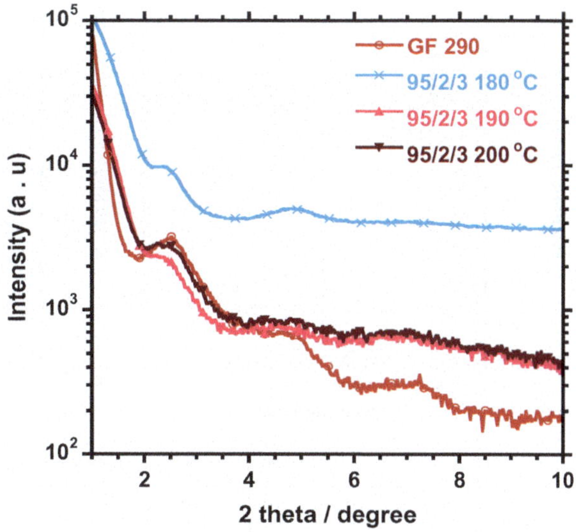

Fig. 5.10 X-ray diffractograms of the nanocomposites prepared via MB method using different temperatures in the extruder

Table 5.7 *d*-spacing of PNCs prepared with different temperature profiles

Sample name PP/PP-g-MA/Clay	2θ (°)	*d*-spacing (nm)
GF290[a]	2.5	3.5
95/2/3 180 °C	2.5	3.5
95/2/3 190 °C	2.5	3.5
95/2/3 200 °C	2.7	3.3

[a]Organically modified calcium bentonite, Betsopa OM™

The XRD patterns of the neat nanoclay used and nanocomposites prepared by MB method using different temperature profiles in the extruder are shown in Fig. 5.10. The *d*-spacing (d_{001}) of all samples are presented in Table 5.7. It is seen that the nanoclay and 95/2/3 180 °C and 95/2/3 190 °C nanocomposites have the same d_{001}. However, the intensity varies between these two samples owing due to the smaller amount of nanoclay stacks present in 95/2/3 180 °C. However, the 95/2/3 200 °C show d_{001} of 2.7. The peak moves to the higher angle of the graph, suggesting that the nanoclay plaletels collapsed during processing because of the degradation of the surfactant as a result of the inherently poor thermal stability of nanoclay.

Figure 5.11a shows the modulus of neat PP and nanocomposites prepared with different temperatures in the extruder. It was observed that the modulus of neat PP slightly increased with the introduction of nanoclay, but the modulus of the nanocomposites is the same irrespective of the processing temperature. Figure 5.11b shows that the elongation of neat PP decreased slightly with nanoclay introduction at 180 °C, but the nanocomposites processed at higher temperatures (190 and 200 °C) have higher elongation than the nanocomposites processed at

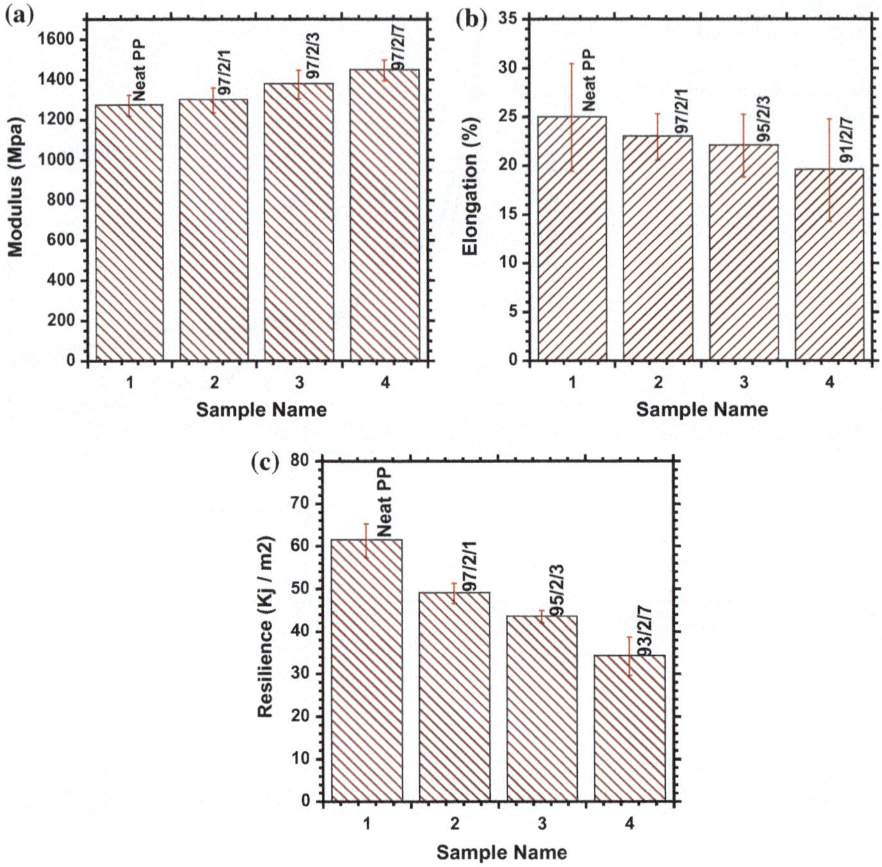

Fig. 5.11 **a** Tensile modulus, **b** elongation at room temperature of neat PP and nanocomposites, and **c** impact strength of neat PP and nanocomposites. The figure displays the average values obtained after ten tests were carried out

180 °C. The elongation of nanocomposites increases as the processing temperature increased. Figure 5.11c show the impact strength of neat PP and nanocomposites produced with different temperatures in the extruder. The impact strength of neat PP is higher than that of nanocomposites. The reduction of impact strength by 59% in the nanocomposites is caused by the incorporation of 3 wt% inorganic nanoclay content in the polymer matrix. Furthermore, the impact strength of nanocomposites is similar in all instances of using different temperatures in the extruder. Table 5.8 shows that neat PP has low HDT compared to the nanocomposites. The addition of nanoclay in the polymer matrix increased the HDT of the nanocomposites. The nanocomposite HDT increases as the processing temperature in the extruder increased. Different temperatures used in the extruder to produce nanocomposites did not affect the softening point of neat PP and nanocomposites.

Table 5.8 HDT-Vicat data for various samples

Sample name (PP/PP-g-MA/Clay)	HDT mean (°C)	Vicat (°C)
100/0/0 (Neat PP)	50 ± 0.67	155 ± 0.3
95/2/3 (180 °C)	54 ± 1.77	154 ± 0.3
95/2/3 (190 °C)	67 ± 4.4	156 ± 0.7
95/2/3 (200 °C)	66 ± 4.0	155 ± 0.5

5.3.4 Protocol 4: Effect of PP-g-MA Content

To study the effect of the PP-g-MA content of PP nanocomposites on the dispersion of nanoclays, different concentrations of PP-g-MA were used, as listed in Table 5.9. The PP nanocomposites were prepared by mixing PP-g-MA and nanoclay in the main hopper to produce an MB of ~31 wt% inorganic nanoclay content, before diluting in PP to 3 wt% nanocomposites. The nanocomposites were produced using SC1. The screw speed and feed rate were kept constant at 202 rpm and 6.6 kg/h, respectively. The temperatures of different zones in the extruder barrel were as follows: 120/160/180/180/180/180/180/180 and 170 °C, respectively, while the die was set at 185 °C.

Figure 5.12 displays the XRD patterns of nanoclay GF290 and nanocomposites prepared by MB method using different concentrations of PP-g-MA. Table 5.10 shows the *d*-spacing (d_{001}) of the same samples. According to this table, the nanoclay and 95/2/3 have the same *d*-spacing of 3.5, 96/1/3 of 3.3, and 92/5/3 of 3.8. The curves in Fig. 5.12 display that the 96/1/3 peak moved to the higher angle of the graph. This is caused by a small quantity of compatibilizer used when producing the nanocomposites, and the graph of the sample is shallow and wide. The 92/5/3 sample moved to the lower angle of the graph owing to the PP-g-MA content used to produce nanocomposites that increased the compatibility between the nanoclay and PP. The sample peak shows intercalation of the polymer chains into the nanoclay platelets, the parallel stacking, and the shallow peak. Hasegawa et al. [11] noticed a similar trend and claimed that to improve the dispersibility of clays in hybrids, modified PP oligomers with larger miscibility to PP should be used so that PP can be inserted between the interlayer of the clays. It is also observed that the dispersion of nanoclay improves with the increase in PP-g-MA content.

Figure 5.13a gives the modulus of the neat PP and nanocomposites. This figure illustrates that the incorporation of nanoclay into the polymer matrix increases the stiffness of the material. However, the increase in PP-g-MA decreases the modulus

Table 5.9 Different content of PP-g-MA in various composites

PP/PP-g-MA-/nanoclay	PP (wt%)	PP-g-MA (wt%)	Nanoclay (wt%)
96/1/3	96	1	3
94/3/3	94	3	3
92/5/3	92	5	3

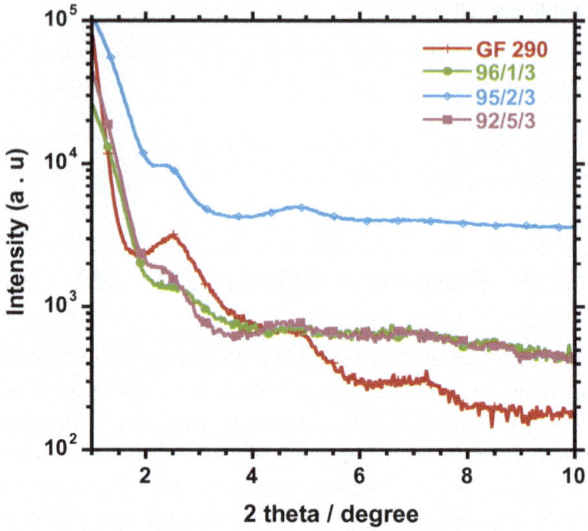

Fig. 5.12 X-ray diffractograms of the nanocomposites prepared via MB method using different content of PP-g-MA

Table 5.10 *d*-spacing of PNCs prepared with different PP-g-MA content

Sample name PP/PP-g-MA/Clay	2θ (°)	*d*-spacing (nm)
GF290[a]	2.6	3.5
96/1/3	2.7	3.3
95/2/3	2.5	3.5
92/5/3	2.3	3.8

[a]Organically modified calcium bentonite, Betsopa OM™

of the polymeric material. The elongation of neat PP and nanocomposites illustrated in Fig. 5.13b shows that the neat PP has higher elongation than the nanocomposites. The 14–25% decrease in elongation is caused by the different concentrations of PP-g-MA used in the extruder when producing nanocomposites. The increase in PP-g-MA content decreases the elongation. Hasengwa et al. [11] observed a similar trend; it was explained that not only the clays but also the PP oligomers contributed to the decline in the elongation of PPCHs (composition of PP, clay, and PP-g-MA). The inclusion of the PP-g-MA caused the composite to fail in a brittle manner as compared to neat PP, causing the decrease of the elongation at break of the nanocomposite as compared to the neat polymer [12]. The HDT of the nanocomposites shown in Table 5.11 is higher than the neat polymer and the nanocomposites.

To summarize phase 4, it was shown that samples prepared using high PP-g-MA content showed better dispersion of nanoclay platelets than samples that used low PP-g-MA content. This behaviour it is attributed to the interaction created by the compatibilizer between the polymer and nanoclay, leading to better bonding effects. Higher PP-g-MA content has enough maleic anhydride functional groups to

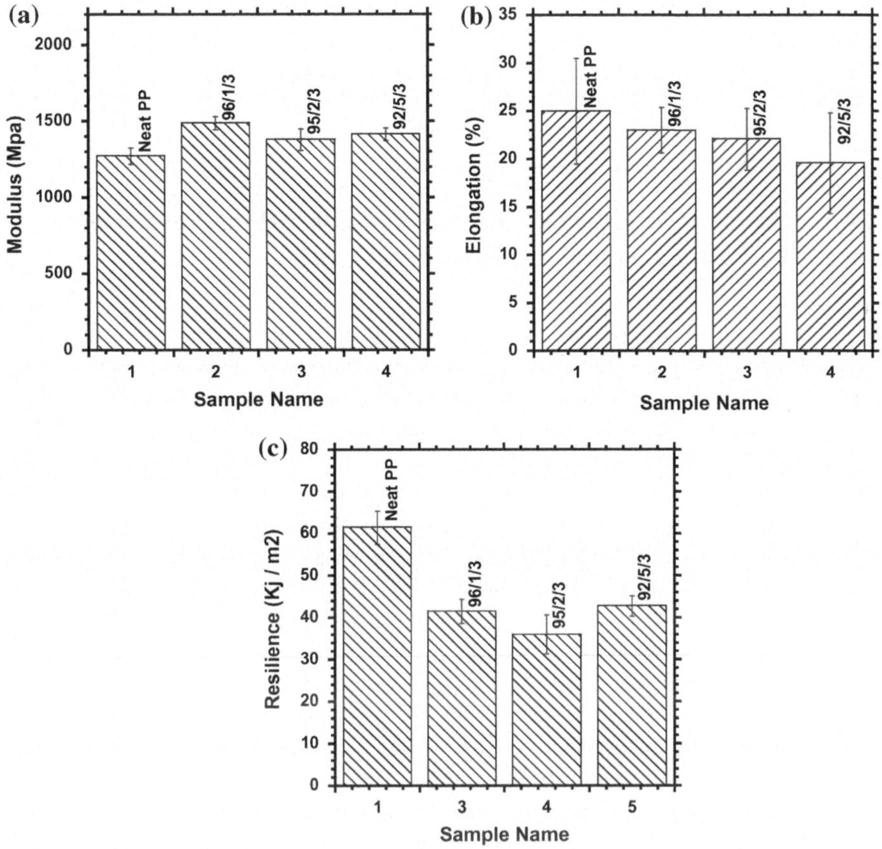

Fig. 5.13 Tensile modulus, elongation, and impact strength at room temperature of neat PP and nanocomposites. The figure displays the average values obtained after ten tests were carried out using different concentrations of PP-g-MA in the extruder

Table 5.11 HDT-Vicat data for various samples

Sample name	HDT mean (°C)	Vicat (°C)
Neat PP	50 ± 0.67	155 ± 0.3
96/1/3	65 ± 7.0	156 ± 0.4
95/2/3	54 ± 1.8	154 ± 0.3
92/5/3	65 ± 4..4	155 ± 0.6

accommodate the bonding side of PP owing to increased carbonyl groups available in the PP-g-MA structure. The main challenge of using high PP-g-MA content is that it is not practical because of its high cost and it is also harmful to environment during processing. On the other hand, the low maleic anhydride content impedes on the catalytic dispersion mechanism effect; thus, phase separation of the materials occurs. However, miscibility is one of the most important factors in the homogeneous dispersion of the nanoclay. Therefore, average content of PP-g-MA was chosen (94/3/3), which has better dispersion of nanoclay and will be viable for industry.

5.3.5 *Protocol 5: Effect of Silicate Concentration Feed Rate and Speed Screw on Residence Time*

To study the effect of feed rate and screw speed on residence time, two additional concentrations were investigated: low concentration of 1 wt% and high concentration of 7 wt% inorganic nanoclay content. PP nanocomposites were obtained by prior preparation of an MB sample by mixing nanoclay and PP-g-MA in a TSE to produce an MB sample with a high concentration of ~31 wt% inorganic nanoclay. Then, the extrudates were dried in a vacuum oven for 24 h at 80 °C. Later, the MB sample was diluted with PP into various concentrations of nanoclay, i.e. 1, 3, and 7 wt% inorganic nanoclay content. The processing temperatures of the extruder barrel used were 120/160/180/180/180/180/180/180/170 °C for zones 2–10, respectively, while the die was set at 185 °C. The screw speeds used were 101, 202, 303, and 396 rpm, while the feed rates were 2.2, 4.4, 6.6, and 10.2 kg/h to process the nanocomposites; one parameter was kept constant and the other parameters were varied. The samples were named 95/2/1, 95/2/3, and 91/2/7. In the presence of ultra-blue tracer, the material will change colour to blue until it reaches the plateau, and then changes back to the original colour. The TSE was allowed to stabilize for 5 min after changing operating conditions before another measurement could start. The video was recorded immediately when 0.25 g of ultra-blue tracer was introduced into the feeder. Images from the video where captured at intervals of 5 s. Then, the image was transferred to image J to determine the intensity. The nanocomposites were prepared using SC1.

After extrusion, all samples were dried overnight in a vacuum oven at 80 °C to remove moisture. Dried extrudates were then injection moulded into test specimens using different moulds, such as a dog bone, DMA, impact, and tensile for mechanical, rheological, and morphological tests using the Engel E-mac 50 injection moulding machine. The profile temperatures of the injection moulding machine were set at 36/220/230/235/240 °C, respectively, while the mould was set at 17 °C.

5.3.5.1 The Effect of Nanoclay Content on Residence Time and Dispersion

Figure 5.14 shows samples produced using a screw speed of 202 rpm, feed rate of 6.6 kg/h, and the concentration of nanoclay was varied as 1, 3, and 7 wt% inorganic

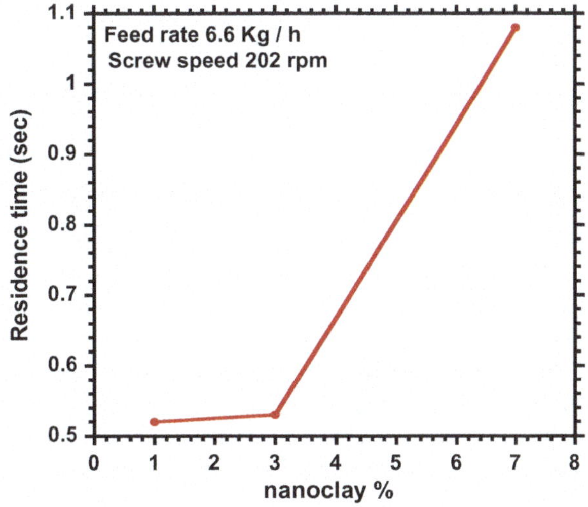

Fig. 5.14 Relates the residence time of different nanoclay concentrations used to produce nanocomposites

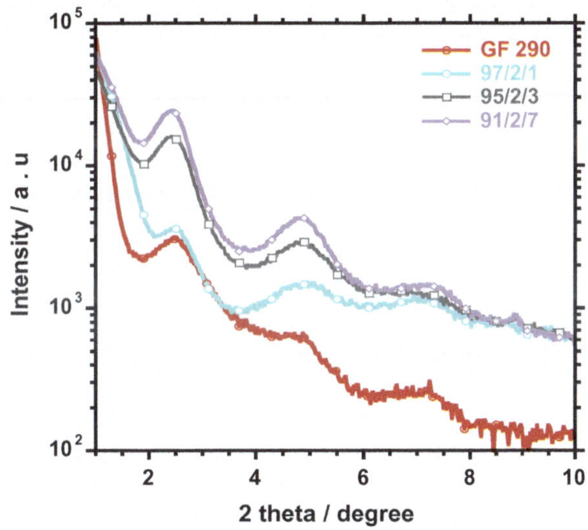

Fig. 5.15 X-ray diffractograms of the nanocomposites prepared via MB method using different concentration of nanoclay

content. This figure suggests that the residence time increases with the increase in nanoclay content. This is because the addition of extra nanoclay to the polymer matrix reduced the viscosity of the nanocomposite, which led to lower residence time.

Figure 5.15 displays the XRD patterns of neat nanoclay used and nanocomposites prepared by MB method using different concentrations of PP-g-MA. Table 5.12 shows the *d*-spacing (d_{001}) of virgin nanoclay used and nanocomposites. According to the table, all nanocomposites and nanoclay GF290 exhibit the same

Table 5.12 *d*-spacing of different nanoclay content composites

Sample name	2θ (°)	*d*-spacing (nm)
GF290	2.5	3.5
97/2/1	2.5	3.5
95/2/3	2.5	3.5
91/2/7	2.6	3.4

d-spacing. However, 91/2/7 has bigger stacks than the 95/2/3, which could be attributed to the large content of nanoclay used when producing nanocomposites. This implies that at low nanoclay concentration the number/amount of nanoclay stacks is lower than at high concentration.

The tensile modulus, elongation at break, and impact strength of neat PP and PP-nanoclay composites are shown in parts (a)–(c) of Fig. 5.16. The neat PP has higher elongation and impact strength. However, the elongation at break and impact strength of nanocomposites decreased when the nanoclay was incorporated into the PP matrix. The elongation of nanocomposites decreased as the percentage of nanoclay increased. Ojijo et al. [13] reported a similar trend using a PLA/PBSA system and showed that this could be due to the poorer adhesion between the polymer matrix and nanoclay. Similarly, the impact strength of the nanocomposites also decreased as the nanoclay loading was increased. This could be because of the agglomerates of the nanoclay that formed during extrusion as supported by XRD in Fig. 5.15. The neat PP has lower tensile modulus in contrast with the nanocomposites. The modulus of the nanocomposites increased with increased nanoclay loading in the PP matrix. This might be because the nanoclay acted as reinforcement to the PP matrix.

5.3.5.2 The Effect of Feed Rate on Residence Time and Dispersion Using 3 wt% Inorganic Nanoclay Content

Figure 5.17 relates the residence time to feed rate using different feed rates of 2.2, 4.4, 6.6, and 10.2 kg/h at a constant screw speed of 202 rpm. This figure illustrates that when the feed rate increases the residence time also decreases. Since the extruder was starve-fed, lower feed rates result in lower degrees of fill, and hence higher residence time. Poulesquen and Vergnes [14] and Bigio et al. [15] reported comparable tendencies, with longer residence times achieved by feeding at low rates owing to partially filled regions of the screw in the extruder by feeding small amounts of material.

Figure 5.18 shows the XRD patterns of the nanocomposites prepared using four different degrees of fill at a constant screw speed of 202 rpm. Table 5.13 shows the *d*-spacing ($d_{(001)}$) of virgin nanoclay used and nanocomposites. The dispersion of nanoclay platelets was not affected by the feed rates

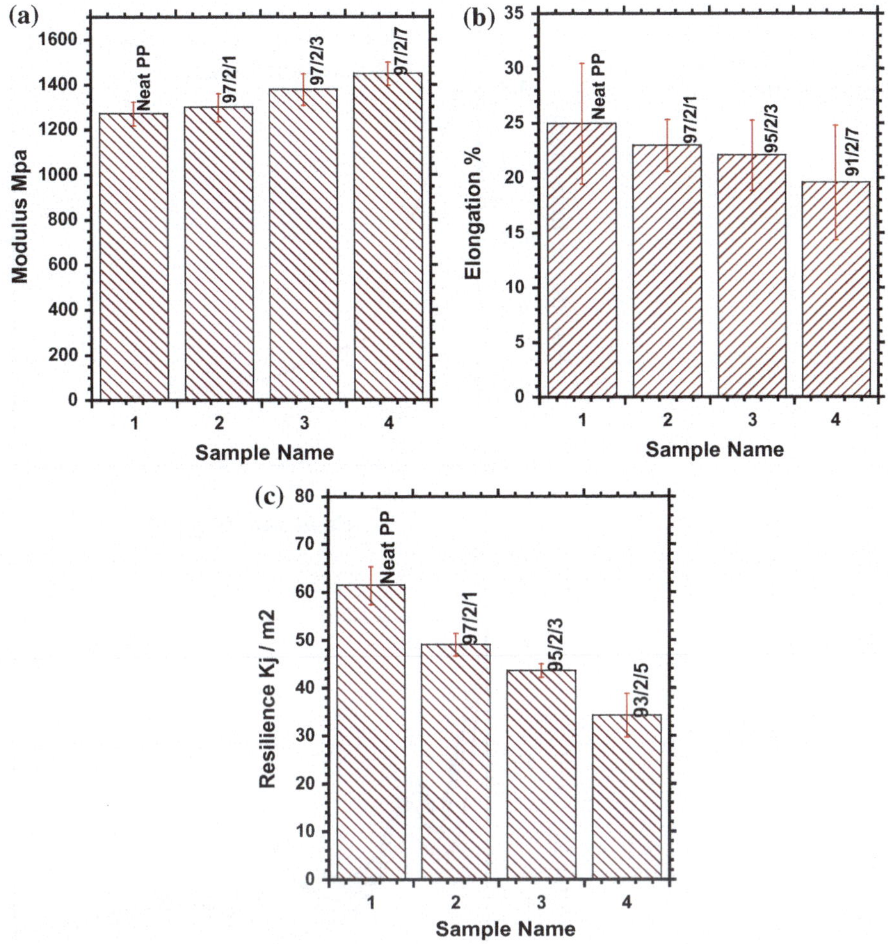

Fig. 5.16 **a** Tensile modulus, **b** elongation, and **c** impact strength at room temperature of neat PP and nanocomposites. The figure displays the average values obtained after ten tests were carried out using different concentrations of PP-g-MA in the extruder

5.3.5.3 The Effect of Screw Speed on Residence Time and Dispersion

Figure 5.19 compares the different screw speeds used in the extruder when producing nanocomposites using a fixed feed rate of 6.6 kg/h. It was observed that the residence time increases as the screw speed increased owing to the rate at which the material is pushed inside the extruder. Bigio et al. [15] and Fel et al. [16] reported a similar trend. The feed rate of 6.6 kg/h was chosen because of the pressure needed for the material flow, short residence time, and less degradation of the material.

The XRD analysis (Fig. 5.20) showed that diffraction peaks linked to the *d*-spacing of the nanoclay stayed the same in all nanocomposites and the neat

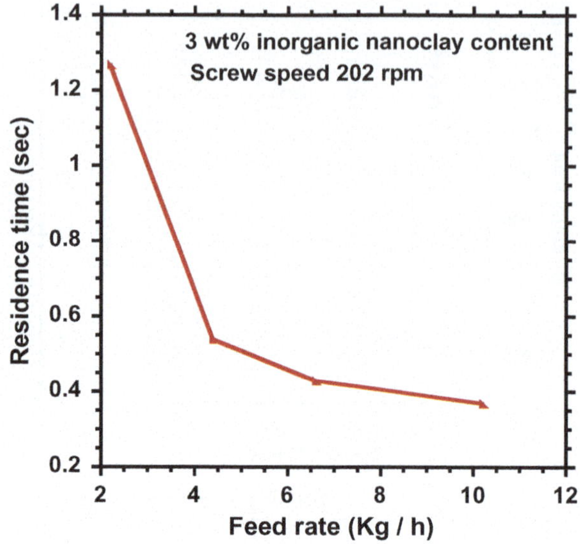

Fig. 5.17 Relates the residence time of diffrerent feed rates used to produce nanocomposites

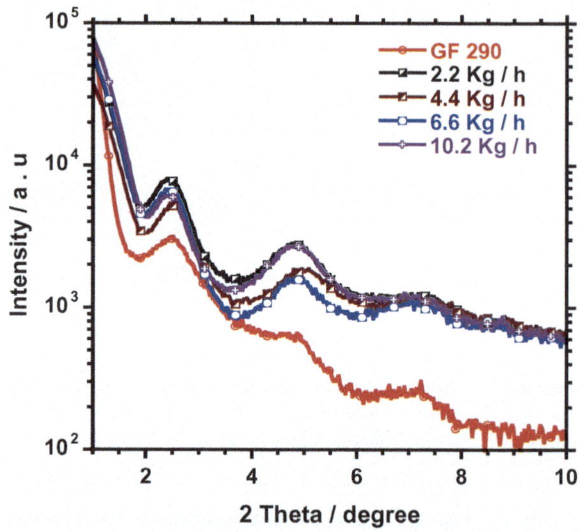

Fig. 5.18 X-ray diffractograms of the nanocomposites produced using different feeding rates and virgin nanoclay

nanoclay with a *d*-spacing of 3.5, regardless the increase in screw speed (Table 5.14). A small decrease in *d*-spacing was observed for the nanocomposites processed at 396 rpm. This could be attributed to the degradation of the nanoclay interlayer surfactant, which is known to promote interlayer spacing reduction [17, 18]. This observation means that intercalation has occurred. The screw speed of 101 rpm is the minimum screw speed to be used in this particular screw since increased screw speed leads to the same dispersion of nanoclay platelets.

Table 5.13 *d*-spacing of different feed rate

Sample name	2θ (°)	*d*-spacing (nm)
GF290[a]	2.5	3.5
2.2 kg/h	2.5	3.5
4.4 kg/h	2.5	3.5
6.6 kg/h	2.6	3.4
10.2 kg/h	2.7	3.3

[a]Organically modified calcium bentonite, Betsopa OM™

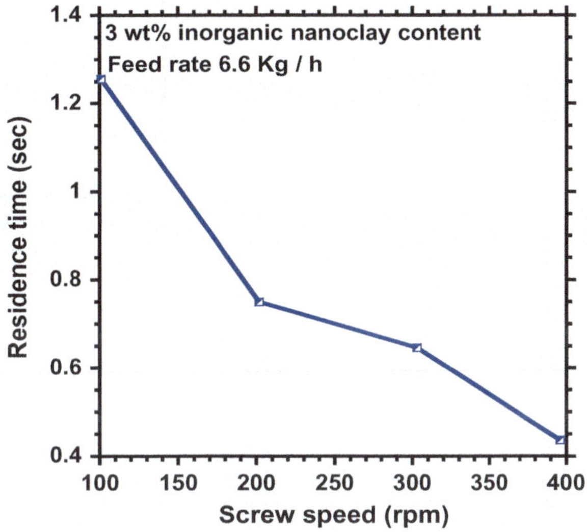

Fig. 5.19 Relates the residence time in the extruder

Irrespective of the screw speed, the dispersion of nanoclay remains the same in the polymer matrix.

To summarize phase 5, PP nanocomposites with different nanoclay contents were prepared in the extruder using a fixed feed rate and screw speed, resulting in nanocomposites with different properties. The nanoclay concentration influenced the properties of the nanocomposites. At high concentration of nanoclay, the XRD-patterns shifted to the higher 2 theta angle (2.6°), indicating the presence of nanoclay agglomeration. As a result, lower impact and elongation were achieved, particularly for 91/2/7. Interestingly, nanocomposite (95/2/3) gave significant increase of modulus and balance impact strength. Also, the RTD characteristics of different feeding rates and screw speeds were investigated using the 95/2/3 sample. It was observed that the increase in feed rate and screw speed influenced residence time. Thus, an increase in feed rate and screw speed significantly decreased the residence time.

Based on the results obtained for all phases, it was found that it is best to produce PP nanocomposites using the MB method with a screw speed of 202 rpm and feed rate of 6.6 kg/h in order to achieve intercalated nanoclay platelets in the polymer matrix.

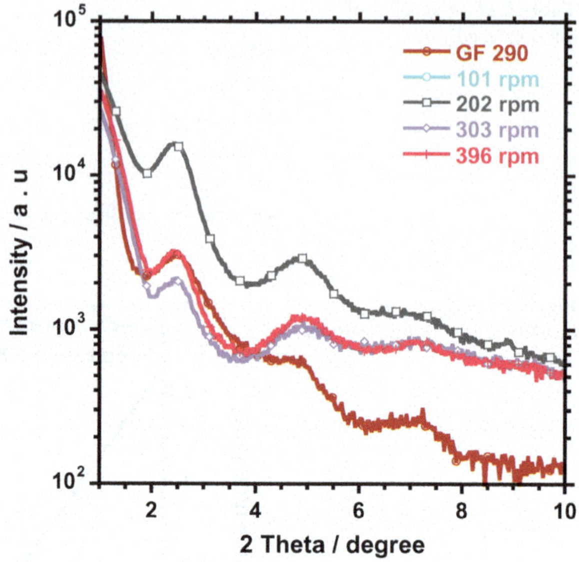

Fig. 5.20 X-ray diffractograms of the nanocomposites and virgin nanoclay

Table 5.14 *d*-spacing of different screw speeds

Sample name	2θ (°)	*d*-spacing (nm)
GF 290[a]	2.5	3.5
101 rpm	2.5	3.5
202 rpm	2.5	3.5
303 rpm	2.5	3.5
396 rpm	2.6	3.4

[a]Organically modified calcium bentonite, Betsopa OM™

5.4 Conclusions

The main aim of this study was to conduct an in-depth investigation of the aspects of processing conditions, such as temperature profile, feed point, screw speed, feed rate, and screw element configuration; the relationship between the various parameters on the dispersion of nanoclays in a PP matrix; and the effects of incorporating nanoclay and loading of compatibilizer on the dispersion of nanoclay in the PP nanocomposite. From the experiments carried out in this work, the following conclusions were drawn based on the objectives of this study.

The PP nanocomposites were successfully produced using MB and SP methods. The use of PP-g-MA to produce a MB sample gave better dispersion of the nanoclay platelets in the polymer matrix. The XRD, SAXS, and TEM results confirmed the improved delamination of nanoclay platelets in the polymer matrix. The sample where nanoclay was introduced in the main hopper showed better elongation at the break of the nanocomposites owing to the few nanoclay agglomerates present in the matrix.

The second objective was to examine the screw element configuration design and its role in the dispersion of nanoclay in PP. There was no significant difference between PNCs made in SC1 and SC2, implying that SC1 offers the threshold shear stress beyond which no improvement in nanoclay dispersion was observed.

The investigation of the various temperature profiles revealed that at high processing temperature, PNC samples exhibited low viscosity and thermal stability. This is due to the fact that at high temperature, the nanoclay platelets collapse, leading to low *d*-spacing and thus poor dispersion. The collapse of the nanoclay was due to the poor surfactant thermal stability, leading to nanoclay platelet agglomeration; as a result, the PNCs exhibited poor thermal stability. On the other hand, samples prepared at low temperature exhibited exclusive viscosity and thermal stability characteristics. Therefore, 180 °C is the optimum processing temperature for PP nanocomposites.

The objective of this phase was to study the influence of PP-g-MA content during compounding of PP nanocomposites on the dispersion of nanoclay. It was observed that samples prepared using high PP-g-MA content showed better dispersion of the nanoclay platelets than samples with low PP-g-MA content. This behaviour it is attributed to the interaction created by the compatibilizer between the polymer and nanoclay, leading to better bonding effects.

The objective of this phase was to investigate the effect of silicate concentration, feed rate, and speed screw on residence time. As the nanoclay content increased at a fixed feed rate and screw speed, the screw speed also increased. An increase in feed rate led to a decrease in residence time. It was concluded that the average feed rate must be used so that the material does not degrade inside the barrel.

Acknowledgements The authors would like to thank the Department of Science and Technology and the Council for Scientific and Industrial Research, South Africa, for financial support.

References

1. Technical Brief: Particle Science, 3. 2011. http://www.particlesciences.com/docs/technical_briefs/TB_2011_3.pdf.
2. Bandyopadhyay J, Ray SS, Scriba M, Wesley-Smith J. A combined experimental and theoretical approach to establish the relationship between shear force and clay platelet delamination in melt-processed polypropylene nanocomposites. Polymer. 2014;55:2233–45.
3. Borse NK, Kamal MR. Estimation of stresses required for exfoliation of clay particles in polymer nanocomposites. Polym Eng Sci. 2009;49:641–50.
4. Fornes TD, Paul DR. Structure and properties of nanocomposites based on nylon-11 and -12 compared with those based on nylon-6. Macromolecules. 2004;37:7698–709.
5. Yuan Q, Misra RDK. Impact fracture behaviour of clay-reinforced polypropylene nanocomposites. Polymer. 2006;47:4421–33.
6. Li J, Ton-That MT, Leelapornpisit W, Utracki LA. Melt compounding of polypropylene-based clay nanocomposites. Polym Eng Sci. 2007;47:1447–58.
7. Lertwimolnum W, Vergnes B. Influence of screw profile and extrusion conditions on the microstructure of polypropylene/organoclay nanocomposites. Polym Eng Sci. 2007;47:2100–9.

8. Nalini R, Nagarajan S, Reddy BSR. Polypropylene-blended organoclay nanocomposites—preparation, characterisation and properties. J Exper Nanosci. 2013;8:480–92.
9. Dong Y, Bhattacharyya D. Experimental characterisation and object-oriented finite element modelling of polypropylene/organoclay nanocomposites. Compos A. 2008;39:1177–91.
10. Santos SK, Demori R, Mauler SR. The influence of screw configurations and feed mode on the dispersion of organoclay on PP. Polimeros. 2013;23:175–81.
11. Hasegawa N, Kawasumi M, Kato M, Usuki A, Akane O. Preparation and mechanical properties of polypropylene-clay hybrids using a maleic anhydride-modified polypropylene oligomer. J Appl Polym Sci. 1998;67:87–92.
12. Ray SS. Clay-containing polymer nanocomposites: from fundamentals to real applications. 1st ed. Oxford: Elsevier; 2013.
13. Ojijo V, Ray SS, Saduku R. Effect of nanoclay loading on the thermal and mechanical properties of biodegradable polyactide/poly[(butylene succinate)-co-adipate] blend composites. ACS Appl Mater Interfaces. 2004;4:2395–405.
14. Poulesquen A, Vergnes B. A study of residence time distribution in co-rotating twin-screw extruders. Part I: theoretical modelling. Polym Eng Sci. 2003;43:1841–8.
15. Bigio DI, Elkouss P, Wetzel MD, Raghavan SR. Influence of polymer viscosity on the residence distributions of extruders. Aiche. 2006;52:1451–9.
16. Fel E, Massardier V, Melis F, Vergnes B, Cassagnau P. Residence time distribution in a high shear twin screw extruder. Int Polym Proc. 2014;29:71–80.
17. Domenech T, Peuvrel-Disdier E, Vergnes B. Influence of twin-screw processing conditions on structure and properties of polypropylene-organoclay nanocomposites. World Congr Polym Proc Soc. 2012;27:517–26.
18. Bandyopadhyay J, Malwela T, Ray SS. Study of the change in dispersion and orientation of clay platelets in a polymer nanocomposite during tensile test by variostage small-angle X-ray scattering. Polymer. 2012;53:1747–59.

Index

S. Sinha Ray (ed.), *Processing of Polymer-based Nanocomposites*,
Springer Series in Materials Science 277,
https://doi.org/10.1007/978-3-319-97779-9

The manufacturer's authorised representative in the EU is Springer Nature Customer Service Centre GmbH, Europaplatz 3, 69115 Heidelberg, Germany. If you have any concerns regarding our products, please contact ProductSafety@springernature.com

Printed and bound by CPI Group (UK) Ltd, Croydon, CR0 4YY
12/08/2026
02194540-0002